D0936644

PROTEIN NUTRITION

PROTEIN NUTRITION

By

HENRY BROWN, A.B., M.D.

Assistant Clinical Professor of Surgery
Harvard Medical School
Assistant Director, Harvard Surgical Service
Boston City Hospital
Surgeon, Massachusetts Institute of Technology

CHARLES C THOMAS • PUBLISHER
Springfield • Illinois • U. S. A.

Published and Distributed Throughout the World by
CHARLES C THOMAS • PUBLISHER
Bannerstone House
301–327 East Lawrence Avenue, Springfield, Illinois, U.S.A.

ISBN 0-398-02897-4

Library of Congress Catalog Card Number: 74-7586

With THOMAS BOOKS careful attention is given to all details of manufacturing and design. It is the Publisher's desire to present books that are satisfactory as to their physical qualities and artistic possibilities and appropriate for their particular use. THOMAS BOOKS will be true to those laws of quality that assure a good name and good will.

Printed in the United States of America

CC-11

Library of Congress Cataloging in Publication Data

Main entry under title:

Protein nutrition.

Papers presented at a conference held at the Boston City Hospital on Nov. 19, 1971; organized by H. Brown and sponsored by the Harvard Medical School Departments of Surgery and Medicine.
1. Protein metabolism—Congresses. 2. Nutrition—Congresses. I. Brown, Henry, 1920– ed. II. Harvard University. Medical School. Dept of Surgery. III. Harvard University. Medical School. Dept. of Medicine. [DNLM: 1. Proteins—Metabolism. QU55 B878p 1973]
QP551.P6973 612'.398 73-7586
ISBN 0-398-02897-4

CONTRIBUTORS

JOHANNES MEIENHOFER, Ph.D.
Lecturer in Biochemistry, Harvard Medical School
Director, Laboratory of Peptide and Protein Chemistry
Children's Hospital, Cancer Research Foundation
Boston, Massachusetts

HENRY BROWN, M.D.
Assistant Clinical Professor of Surgery, Harvard Medical School
Assistant Director, Harvard Surgical Service, Boston City Hospital
Surgeon, Massachusetts Institute of Technology
Boston, Massachusetts

JULIE BROWN, A.B., M.S.
Research Assistant, Children's Hospital
Cancer Research Foundation; Harvard Surgical Service, Boston City Hospital
Boston, Massachusetts

STANLEY M. LEVENSON, M.D.
Professor of Surgery
Albert Einstein College of Medicine of Yeshiva University
New York, New York

ELI SEIFTER, Ph.D.
Department of Surgery
Albert Einstein College of Medicine of Yeshiva University
New York, New York

CHARLES S. DAVIDSON, M.D.
Professor of Medicine, Harvard Medical School
Director of Gastroenterology, The Second and Fourth Surgical Services
Thorndike Memorial Laboratories, Boston City Hospital
Boston, Massachusetts

ALFRED E. HARPER, Ph.D.
Professor of Biochemistry
Department of Nutritional Sciences and Biochemistry
University of Wisconsin
Madison, Wisconsin

T. T. AOKI, M.D.
Instructor of Medicine, Harvard Medical School
Joslin Diabetes Foundation
Peter Bent Brigham Hospital
Boston, Massachusetts

MURRY F. BRENNAN, M.D.
Fellow in Surgery, Harvard Medical School, Department of Surgery
Peter Bent Brigham Hospital
Boston, Massachusetts

W. A. MULLER
Swiss National Foundation
Research Fellow in Medicine
The Harvard Medical School
Boston, Massachusetts

GEORGE F. CAHILL, JR.
Professor of Medicine
The Harvard Medical School
Director of Research
Joslin Diabetes Foundation
Boston, Massachusetts

ROBERT L. RUBERG, M.D.
Instructor in Surgery, University of Pennsylvania School of Medicine
Philadelphia, Pennsylvania

EZRA STEIGER, M.D.
Instructor in Surgery, University of Pennsylvania School of Medicine
Philadelphia, Pennsylvania

CHARLES VAN BUREN, M.D.
Harrison Department of Surgical Research
University of Pennsylvania School of Medicine
Philadelphia, Pennsylvania

STANLEY J. DUDRICK, M.D.

Professor and Director, Program in Surgery
University of Texas Medical School at Houston
Houston, Texas

GEORGE L. BLACKBURN, M.D., Ph.D.

Assistant Professor of Surgery
The Harvard Medical School
Senior Research Associate
Massachusetts Institute of Technology
Boston, Massachusetts

JEAN-PIERRE FLATT, Ph.D.

Professor of Biochemistry
University of Massachusetts
School of Medicine
Worcester, Massachusetts

INTRODUCTION

Adequate protein nutrition is difficult to define, particularly in disease, because means of fixing nitrogen are not well understood. Further, few mechanisms for joining nitrogen compounds, in this instance amino acids, are known.

Protein synthesis in each of us in health proceeds rapidly and smoothly under relatively mild conditions. In contrast, to duplicate such synthesis in the laboratory, very rigorous conditions with harsh reagents, including very concentrated and corrosive acid solutions are usually required. This dichotomy is being resolved by the protein chemist, who is constantly improving synthetic methods. His work, of first order importance, will not only give us new insight into mechanisms for new protein formation, but also synthetic peptides and protein models for nutritional investigation.

We know, for example, that most patients use certain whole protein parenterally very well, as albumin or all plasma protein. Other patients will use proper mixtures of amino acids either orally or intravenously so well that they need no other nitrogen source. In many illnesses, on the other hand, amino acids are not well accepted by the patient. Many reasons have been advanced, with merit to most of them. The patient may have an endocrine imbalance as in diabetes or shock preventing proper nitrogen use. Some peptide bonds critical for protein synthesis may be difficult for the ill patient to make. Ratios of amino acids given may be unsuitable for a particular patient, particularly since large concentrations of single amino acids may be toxic. Other needed factors may be missing from the diet, or nutrients from gut bacteria, analogous to vitamin K, may be lacking.

For these and many other reasons I have requested authors of essays for this book, each an expert in his field, to meet here to discuss these problems. From such meetings will come new approaches and better treatment for our patients, the ultimate goal of all clinical research.

ACKNOWLEDGMENT

Thanks are due to each of the essayists and particularly to Dr. Charles Davidson for his help in arranging the Conference. Our appreciation is also expressed to Boston City Hospital, the place of the Conference, and to the Departments of Surgery and Medicine of the Harvard Medical School under whose auspices the meeting was held.

CONTENTS

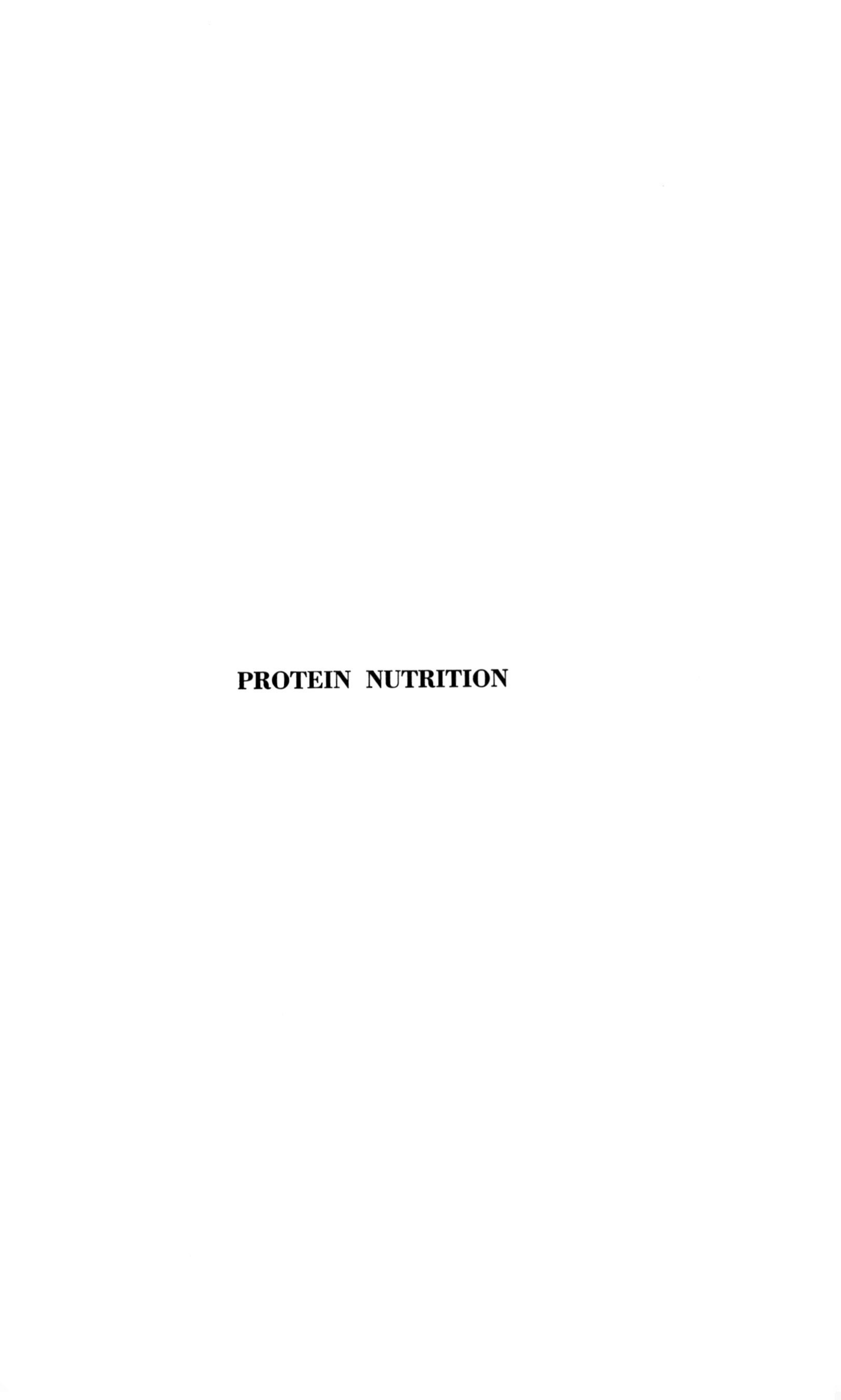

PROTEIN NUTRITION

CHAPTER 1

CHEMICAL ASPECTS OF PEPTIDE AND PROTEIN SYNTHESIS(1)

JOHANNES MEIENHOFER

INTRODUCTION

INVOLVEMENT OF PEPTIDE chemistry in research on protein nutrition would not appear to be obvious at first sight except, perhaps, for a historic case. The famous chemist, Emil Fischer, father of peptide synthesis, was presented by a student with a synthetic peptide which was the result of much work. Fischer closely inspected the product and, finally, tasted it. He found it to be of good flavor—so good, indeed, that in another instant the entire sample of the unhappy student was consumed.

In modern biochemical research peptide and protein chemists have been concerned with mechanisms of enzyme catalysis or hormone action. During the past 15 years, methodological progress has allowed chemists to prepare by conventional procedures peptides with chain lengths of up to approximately 40 amino acids. In the cell, such peptides are synthesized within seconds. Experienced chemists, however, might need months or even years to accomplish the same, and it seems very uneconomical to try to compete with nature in this manner. This unfavorable relation changes at once when one considers (2) that the completion of a gene mutation involving the replacement of a single amino acid in a protein such as hemoglobin takes 7 to 10 million years.

In contrast, chemical synthesis can produce series of peptide analogs of a given system within reasonable time.

Thus peptide synthesis allows the study of structure-activity relationships. Another incentive to synthesis comes from the scarce natural supply of some highly-active peptides.

Decisive impetus and direction in the chemical synthesis of biologically-active peptides came from du Vigneaud's first synthesis (3) of oxytocin (Fig. 1-1) in 1953. A growing number of efficient synthetic methods (4,5) accelerated the progress. And recently, several reports on chemical syntheses of proteins such as ribonuclease (6–8) or human growth hormone (9) have attracted wide attention. At the same time, several highly potent hormonal control factors were found to be surprisingly small peptides. Both, thyrotropin releasing hormone, TRH, and melanotropin-release inhibiting

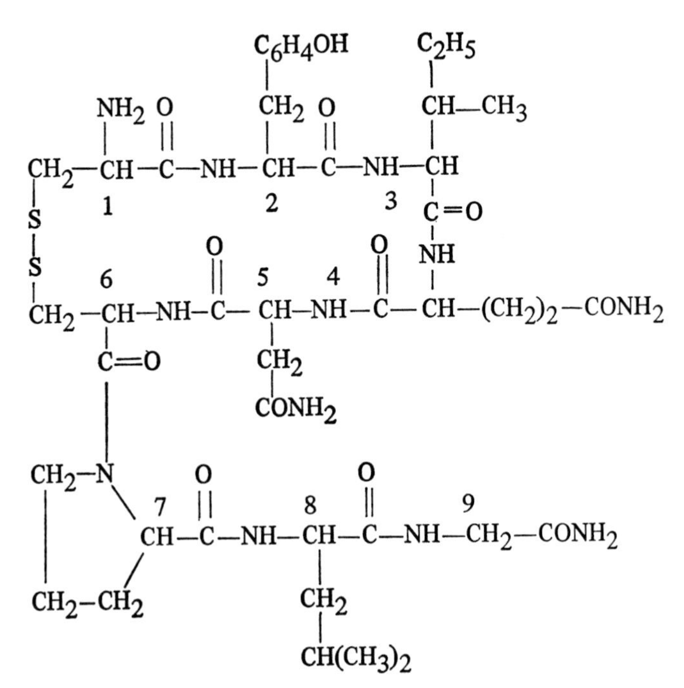

Figure 1-1. Structure of oxytocin. Numbers indicate positions of component amino acid residues.

factor are tripeptides. They occur in the producing glands in such minute quantities (nanomole-level) that it required peptide synthesis to come to the aid of analytical studies in establishing their structure. Thus, TRH is L-pyroglutamyl-L-histidyl-L-prolinamide (**1**) (10,11) and melanotropin-release inhibiting factor is L-prolyl-L-leucyl-glycinamide (**2**). (12)

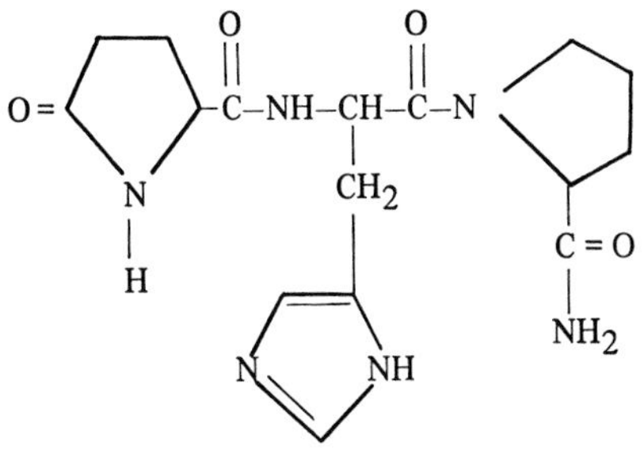

(1) 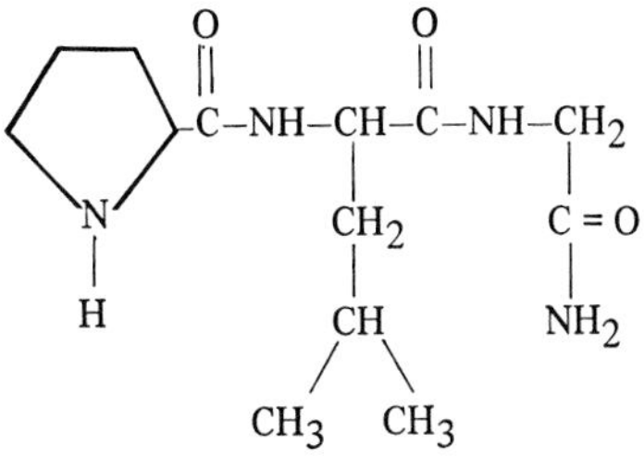 (2)

Peptide hormones that play important roles in protein nutrition include insulin* and glucagon (Fig. 1-2). Chemical syntheses have been achieved of these hormones (13–16) and of others such as gastrin (17) or secretin (18,19) (Fig. 1-2). The search for an active core in gastrin, for example, led to the recognition that the C-terminal tetrapeptide, Trp-Met-Asp-Phe-NH_2, possesses full biological potency. (20) Several hundred analogs of this and related penta- and hexapeptides have been prepared. (21) Approximately 50 analogs and homologs of insulin have thus far been synthesized in efforts to establish structure-activity correlations. (22–24) A total synthesis of glucagon (16) produced 5 grams of crystalline hormone thus, for the first time, making it available for biological research in more than trace amounts. In these ways, modern peptide chemistry is playing an increasingly important role in research on protein nutrition.

Peptide synthesis has traditionally been distinguished by a somewhat confusing multitude of methods and procedures for a seemingly simple matter: the repeated formation of the

* See Chapter 6 for a detailed discussion.

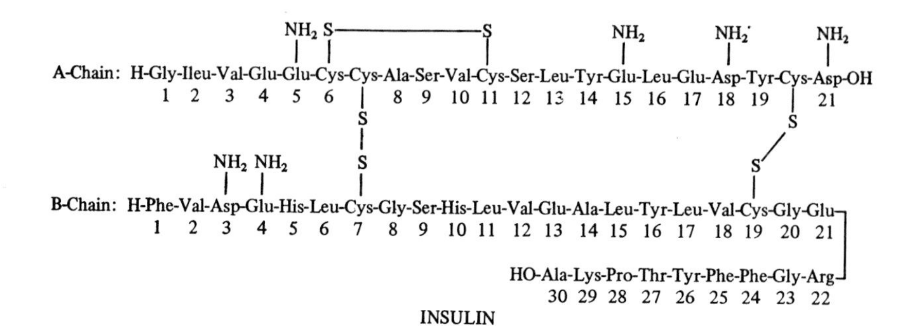

NH_2 (at 3)

His-Ser-Glu-Gly-Thr-Phe-Thr-Ser-Asp-Tyr-Ser-Lys-Tyr-Leu-Asp-Ser-
1 2 3 4 5 6 7 8 9 10 11 12 13 14 15 16

NH_2 (at 20), NH_2 (at 24), NH_2 (at 28)

Arg-Arg-Ala-Glu-Asp-Phe-Val-Glu-Trp- Leu-Met-Asp-Thr
17 18 19 20 21 22 23 24 25 26 27 28 29

GLUCAGON

Glu-Gly-Pro-Trp-Val-Glu-Glu-Glu-Glu-Ala-Ala-Tyr-Gly-Trp-Met-Asp-Phe-NH_2
1 2 3 4 5 6 7 8 9 10 11 12 13 14 15 16 17

GASTRIN

His-Ser-Asp-Gly-Thr-Phe-Thr-Ser-Glu-Leu-Ser-Arg-Leu-Arg-
1 2 3 4 5 6 7 8 9 10 11 12 13 14

NH_2 (at 20), NH_2 (at 24)

Asp-Ser-Ala-Arg-Leu-Glu-Arg-Leu-Leu-Glu-Gly-Leu-Val-NH_2
15 16 17 18 19 20 21 22 23 24 25 26 27

SECRETIN

Figure 1-2. Structures of bovine insulin, bovine glucagon, bovine gastrin, and porcine secretin.

same chemical bond. However, peptides of identical amino acid composition but with varied sequences can differ tremendously in chemical properties. This is important for the known large variety of biological roles of different peptides, but it does not facilitate their chemical synthesis. In addition, the necessity of selective protection and deprotection of functional groups adds to the complexity.

With the development of solid-phase peptide synthesis by Merrifield (25–27) a convenient and rapid procedure became available. Its potential for fast syntheses of biologically-active peptides was shown by preparations of bradykinin and angiotensins in half-gram amounts.

Thus two principal approaches are now available: (a) conventional synthesis in solution, which requires considerable effort, time and experience but generally affords homogeneous products, (b) solid-phase synthesis, which is fast and easy to manipulate but can give products of considerable heterogeneity. In the following discussion both methods will be reviewed and their comparative advantages and present shortcomings evaluated. This might aid in choosing between the two approaches depending on the nature and objective of a given study.

PEPTIDE SYNTHESIS IN SOLUTION

Basic Scheme of Synthesis

Peptides consist of amino acids which are joined together by amide linkages to form molecules which are often very large. (4,5) Peptide bond formation formally results from condensation of amino acids with elimination of water **(3)**.

$$\overset{\oplus}{H_3N}\text{–}CHR^1\text{–}CO\overset{\ominus}{O} + \overset{\oplus}{H_3N}\text{–}CHR^2\text{–}CO\overset{\ominus}{O} \longrightarrow \overset{\oplus}{H_3N}\text{–}CHR^1\text{–}CO\text{–}NH\text{–}CHR^2\text{–}CO\overset{\ominus}{O} + H_2O \quad (3)$$

Chemical synthesis, basically, proceeds through three stages (Fig. 1-3). It begins with preparations of protected amino acids. If reaction **(3)** is to produce a single defined peptide there must be *one* free amino group and *one* free carboxyl group. Functional groups that are not to react must be blocked by protecting groups. Substitution of the α-amino group provides the so-called "**carboxyl component**"; substitution of the α-carboxyl group affords the "**amine component**." Side-chains R may also contain reactive functions which have to be blocked. Two types of protective groups are thus needed, (a) one type that protects the α-amino and the α-carboxyl groups intermittently and that can be selectively cleaved, and (b) another type that blocks functional groups permanently throughout a synthesis. Both must be

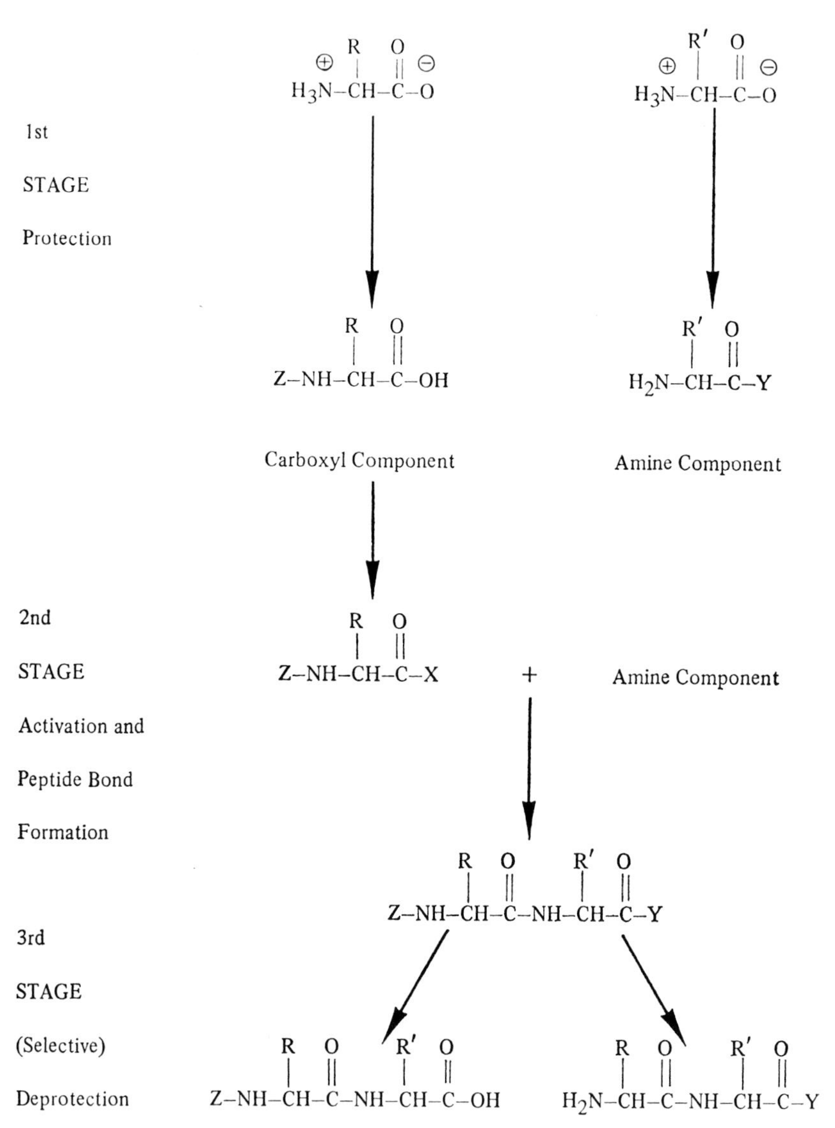

Figure 1-3. General scheme of peptide synthesis. Z, Amine protecting group; Y, carboxyl protecting group; X, activating substituent.

removable at the end of the synthesis under conditions that do not affect the final peptide.

The actual peptide bond formation (second stage in Fig. 1-3) is accomplished by activation of the carboxyl group of the carboxyl component followed by reaction with the amine component to produce a protected peptide. A good condensation method must fulfill five requirements: (28) minimal racemization,* minimal side reactions, easy work-up, high yield, and rapid peptide bond formation. As a consequence of the stringent requirements only four methods are thus far useful enough for general applicability and wide use (see below).

In the third stage (Fig. 1-3) protecting groups are removed as required. Partial selective deprotection affords intermediates for further condensation. Complete deprotection at the end provides the desired free peptides.

Peptide syntheses are, with few exceptions,† carried out by following this basic scheme. Higher peptides are prepared by alternating repetitions of stages 2 and 3.

Protecting Groups

To be useful, a protecting group must be easily introduced into amino acids and peptides, completely stable during peptide bond formation and readily cleaved afterwards without deleterious effects on peptide bonds. High cleavage selectivity is required to achieve complete removal of temporary protecting groups and simultaneous full stability of permanent protecting groups. Selection of the best combination of different protecting groups for α-amino, ω-amino, carboxyl and other functional groups is important when designing a new synthesis.

---

* The common amino acids, except glycine, possess an asymmetric α-carbon atom. In most biologically active peptides they are of the L-configuration. Microbial peptides can also contain D-amino acids, but all peptides of biological origin are optically uniform. Therefore, racemization must be avoided during chemical synthesis.

† See, e.g. Denkewalter *et al.* (29) for the N-carboxyanhydride method of peptide synthesis with defined sequence.

The most commonly used **amino protecting groups** are the benzyloxycarbonyl group **(4)**, (30) the *tert*-butyloxycarbonyl group, abbreviated "Boc" **(5)**, (31) and derivatives of these two. Less frequently used are the *o*-nitrophenylsulfenyl group **(6)**, (32,33) the *p*-toluenesulfonyl group **(7)**, (34,35) the formyl **(8)**, (36–39) the trifluoroacetyl group **(9)**, (40) and the phthalyl group **(10)**. (41,42) Improved pro-

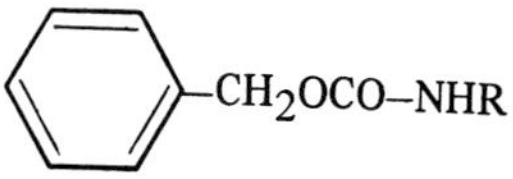

C_6H_5–CH_2OCO–NHR

(4)

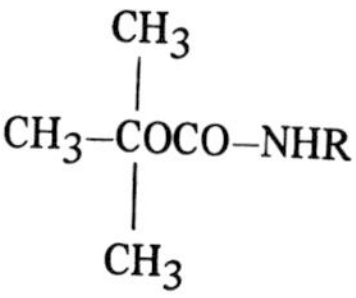

CH_3–C(CH_3)(CH_3)–OCO–NHR

(5)

NO_2–C_6H_4–S–NHR

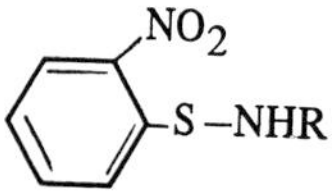

(6)

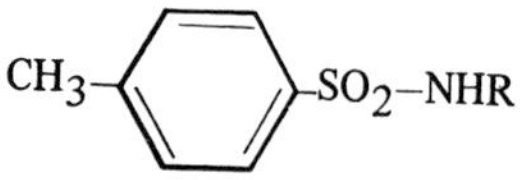

CH_3–C_6H_4–SO_2–NHR

(7)

HCO–NHR

(8)

CF_3CO–NHR

(9)

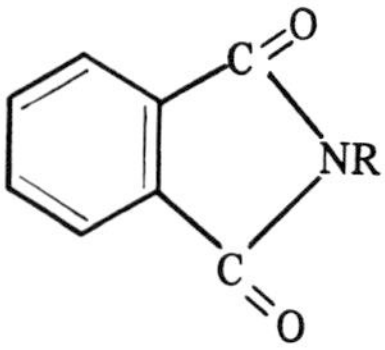

C_6H_4(C=O)(C=O)NR

(10)

cedures for the introduction of some of these protecting groups into amino acids (or peptides) have recently been described. For instance, *tert*-butyloxcarbonyl fluoride **(11)** (43) reacts 5 to 10 times as fast as the commonly used Boc-azide **(12)** (44) greatly facilitating preparation of *tert*-butyloxycarbonylamino acids, especially if used with pH-stat controlled addition of base. (45–47)

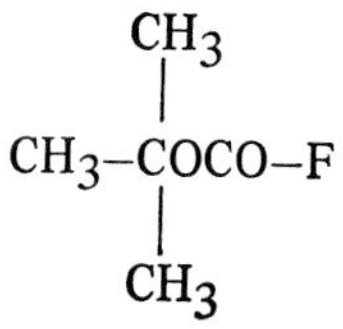

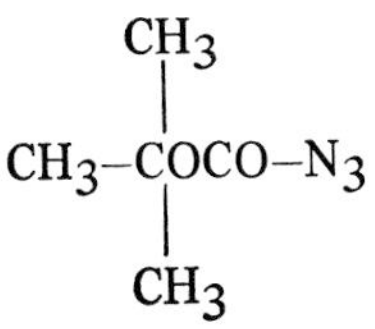

$$CH_3-\overset{\displaystyle CH_3}{\underset{\displaystyle CH_3}{|}}\!\!\!\!\!\!\!\!\overset{|}{C}OCO-F \qquad\qquad CH_3-\overset{\displaystyle CH_3}{\underset{\displaystyle CH_3}{|}}\!\!\!\!\!\!\!\!\overset{|}{C}OCO-N_3$$

(11) (12)

For **protection of carboxyl groups** of amino acids and peptides methyl esters **(13)**, ethyl esters **(14)**, benzyl esters **(15)**, (48) and *tert*-butyl esters **(16)**, (49,50) are most frequently employed. Substituted hydrazides **(17)** (51,52) and the amide function are used as C-protecting groups when

$RCO-OCH_3$ (13) $RCO-OC_2H_5$ (14) $RCO-OCH_2-C_6H_5$ (15)

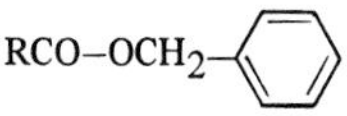

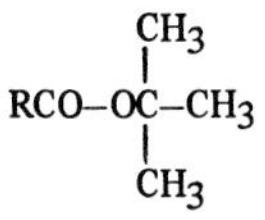

$RCO-OC(CH_3)_2-CH_3$ (16) $RCO-NHNH-R'$ (17) $R' = -COOCH_2-C_6H_5$; $= -COOC(CH_3)_3$

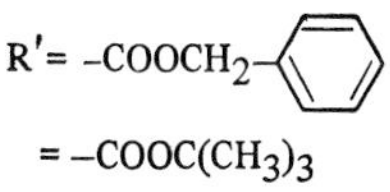

applicable. Commonly used amine and carboxyl protecting groups are listed in Table 1-I with reagents most frequently employed for their introduction and cleavage.

For **protection of** ω-amino and ω-carboxyl functions in **side-chains** of amino acids or peptides, protecting groups mentioned above are generally used. Special substituents are required for the mercapto function of cysteine and sometimes for hydroxyl functions of serine, threonine and tyrosine, and for the imidazole moiety of histidine. Two of the most frequently used protecting groups for these functions are the benzyl substituent, providing S-benzyl cysteine, (62) and the *tert*-butyl substituent for the hydroxyl functions. A novel thiol protecting group is the acid- and base-stabile acetamido-

TABLE

COMMONLY USED AMINE AND CARBOXYL PROTECTING GROUPS

Name	*Abbreviation*	*Reagent for Introduction into Amino Acids*	Acidolysis	
			HF[56,57]	HBr in HOAc / TFA[5
AMINE PROTECTING GROUPS				
Benzyloxycarbonyl	Z	Z–chloride[30]	+	+
tert-Butyloxycarbonyl	Boc	Boc-azide[31,44 c]	+	+
Formyl[d]	For	HCOOH–DCCI[53 e]	–	–
o-Nitrophenylsulfenyl	Nps	Nps-chloride[32,33]	+	+
Phthalyl	Pht	Pht–$NCOOC_2H_5$[54 f]	–	–
p-Toluenesulfonyl	Tos	Tos-chloride[34]	–	–
Trifluoroacetyl	Tfa	Tfa-anhydride[40 g]	–	–
CARBOXYL PROTECTING GROUPS				
Methyl, Ethyl ester	OMe,OEt	H^+–catalyzed esterification	–	–
Benzyl ester	OBzl	Bz1OH–H^+[48]	+	+
tert-Butyl ester	OBu^t	Isobutylene-H^+[49]	+	+

REAGENTS EMPLOYED FOR INTRODUCTION AND CLEAVAGE.[a]

CLEAVAGE						
Acidolysis		Reduction		Base Treatment		
FA[60] or CH_2Cl_2[b]	BF_3–HOAc[61]	H_2–Pd[30]	Na–liq. NH_3[35,62-64]	2N NaOH–acetone[65]	1M Piperi–dine–H_2O[66]	NH_3 or N_2H_4 alcohol[4]
(–)	–	+	+	–	–	–
+	+	–	–	–	–	–
–	–	–	–	–	–	–
+	+	X	X	–	–	–
–	–	–	–	X	X	+
–	–	–	+	–	–	–
–	–	–	X	+	+	–
–	–	–	X	+	–	+[h]
(–)	–	+	+	+	–	+[h]
+	+	–	–	–	–	–

emovable; – stable; X incompatible; (–) incomplete stability (some cleavage).
, hydrogen fluoride; HBr, hydrogen bromide; BF^3, boron trifluoride; N_2H_4, hydrazine.
luoroacetic acid in methylene chloride (1:1). [c]*tert*-Butyloxycarbonyl fluoride may also be used[43].
novable under special conditions of acidolysis, (1N HC1 in alcohol), hydrogenolysis, or oxidation[4,5].
traditional procedure using formic acid in acetic anhydride[39] can give racemization and side product
ation. [f]Phthalic anhydride may also be used[41, 42]. [g]Unsymmetrical trifluoro trichloro acetone affords
oroacetylation under mild neutral conditions[55]. [h]This treatment results in conversion to Peptide amides
eptide hydrazides respectively.

methyl group **(18)**. (67) It is easily cleaved at room temperature by mercuric ions at pH 4. Most importantly, solubility characteristics of the acetamidomethyl group allow its use in both organic solvents and aqueous media.

$$RS—CH_2NHCOCH_3 \quad \textbf{(18)}$$

The guanidino function of arginine is occasionally protected by protonation, but more frequently the nitro substituent (68) is used. A mild and efficient procedure using potassium nitrate in hydrogen fluoride at 0° for preparation of nitroarginine has been described recently. (69)

Methods of Peptide Bond Formation

Four methods of peptide bond formation are sufficiently effective to have found wide and general application. In historical order they are the azide method, (70) the mixed anhydride method, (71–73) the dicyclohexylcarbodiimide method (74) and the active ester method, in particular the *p*-nitrophenyl ester method. (75) Recent progress in the field has primarily been made by improvement, refinement, and expansion of these four methods rather than by development of new methods. The four methods will be briefly described and evaluated on the basis of five experimental criteria. (i) Racemization at the α-carbon atom of those amino acids that participate in the peptide bond formation must be avoided or kept at very low levels (76) ($< 0.1\%$). (ii) Side reactions should be avoided. (iii) Isolation of the peptide and separation from by-products and from contaminating side products should be facile and effective. Most analytical criteria for homogeneity of synthetic peptides are by themselves of limited significance and a number of independent data have to be accumulated. Proof of homogeneity becomes increasingly difficult with growing peptide chain length. (iv) Yield should be high since peptide syntheses are multistage procedures. (v) Time required for one peptide bond formation (including work-up, i.e. product isolation) should not exceed three days.

The *azide method* (70) proceeds through two separate stages. First is the preparation of N-protected amino acid or peptide hydrazides **(20)** by treatment of corresponding methyl **(19)** or ethyl ester with hydrazine in alcohol or dimethylformamide. In the second stage the azide is formed **(21)**. The hydrazide is treated with sodium nitrite in 1*N* aqueous HCl or with butyl nitrite (77) in an organic solvent. Reaction with the amine component affords the desired peptide with liberation of hydrazoic acid **(22)**.

$$\underset{(19)}{RCO\text{–}OCH_3} \xrightarrow{H_2NNH_2} \underset{(20)}{RCO\text{–}NHNH_2} \xrightarrow{HNO_2} \underset{(21)}{RCO\text{–}N_3} \xrightarrow{+\ H\text{–}NHR'} \underset{\text{Peptide}}{R\text{–}CONH\text{–}R'} + \underset{(22)}{HN_3}$$

RCO——: N-protected amino acid or peptide
——NHR′: free or C-protected amino acid or peptide residue

Criteria: (i) Racemization is generally very low or too low to be observed. Recently however, as much as 20 percent racemization was detected after a condensation of peptide intermediates during calcitonin synthesis. (78) (ii) Many side reactions can occur, (79) such as amide formation, or Curtius rearrangement to isocyanates. (iii) Work-up would be easy in case of quantitative reaction because hydrazoic acid **(22)** evolves as a gas. Unreacted starting components are removed by solvent distribution (washing) or other fractionation procedures but side products are often difficult to separate. (iv) Yields are between 30 percent and 70 percent. (v) Time required (including hydrazide formation) ranges from two to six days.

The *mixed anhydride method.* (71–73) In the carbonic carboxylic anhydride **(26)** procedure (80) the carboxyl component **(23)** is dissolved in a suitable anhydrous solvent such as tetrahydrofuran or dimethylformamide. It is then treated at —10° to —15° with one equivalent of isobutyl chloroformate **(24)** in the presence of one equivalent of N-methylmorpholine **(25)**. The ensuing mixed anhydride **(26)**, is then combined with the amine component **(27)**.

Criteria: (i) Racemization can be kept minimal through strict observation of improved reaction conditions elaborated by Anderson *et al.* (80) Excess base, even in trace amounts,

$$RCO{-}OH + ClCOOCH_2CH(CH_3)_2 \xrightarrow{CH_3{-}N\langle\text{morpholine}\rangle} RC(=O){-}O{-}C(=O)OCH_2CH(CH_3)_2$$

(23) (24) (25) (26)

$$\xrightarrow{H{-}NHR'} R{-}CONH{-}R' + HOCH_2CH(CH_3)_2 + CO_2$$

(27) Peptide (28) (29)

can lead to considerable racemization. (ii) Side reactions occur only infrequently, such as diacylimide formation (81,82) when the amine component contains N-terminal glycine. (ii) Work-up is easy because the by-products **(28)** and **(29)** are volatile. (iv) Yields are generally high: 50 percent to > 99 percent. (v) The method is very fast requiring only a few hours to one day.

The *dicyclohexylcarbodiimide method* (74) provides a highly reactive "condensing agent" **(32)**. It is conveniently added at 0° to the mixture of carboxyl **(30)** and amine **(31)** components dissolved in suitable organic solvents such as acetonitrile, ethyl acetate or dimethylformamide. The reaction rapidly forms peptide and N,N'-dicyclohexylurea **(33)** even in the presence of water. (74,83)

$$RCO{-}OH + H{-}NHR' \xrightarrow{\langle H\rangle{-}N{=}C{=}N{-}\langle H\rangle} R{-}CONH{-}R' + \langle H\rangle{-}NHCONH{-}\langle H\rangle$$

(30) (31) (32) Peptide (33)

Criteria: (i) Racemization can be unacceptably high during condensation of peptide intermediates. Significant lowering or suppression of racemization is effected by (α) cooling to 0°, (β) utilizing unpolar solvents, as toluene, (84) provided **(30)** and **(31)** are sufficiently soluble, and (δ) adding N-hydroxysuccinimide **(34)** (85,86) or l-hydroxybenzotriazole **(35)**. (87) (ii) A frequently observed troublesome side reaction is formation of N-acyl-dicyclohexylureas **(36)**. (88) They are difficult to separate from the peptide.

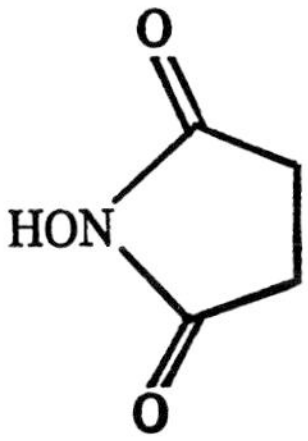

(34) 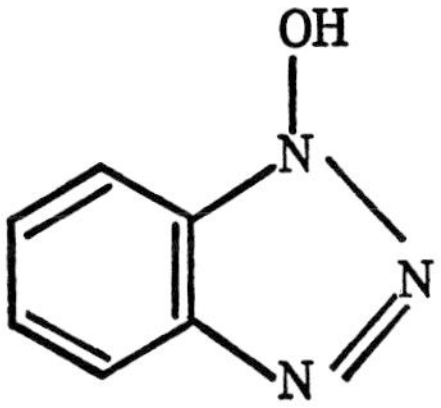 (35)

The side reaction is suppressed by addition of N-hydroxysuccinimide **(34)** (85,86) or l-hydroxybenzotriazole **(35)**. (87) Dehydration of the ω-carbonamide groups of glutamine and asparagine produces undesirable nitriles. (iii) Work-up is sometimes complicated by incomplete separation of the di-

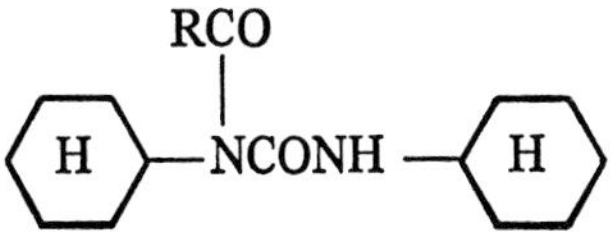

(36)

cyclohexylurea by-product. (iv) Yields range from 30 percent to $>$ 90 percent. (v) One to three days are required to complete a carbodiimide condensation.

The *active ester method* excels in variety. The readily available, stable crystalline *p*-nitrophenyl esters **(37)** (75) of N-protected amino acids are most popular. Other useful active esters include N-hydroxysuccinimide **(38)** (89), 2,4,5-trichlorophenyl **(39)** (90) and pentachlorophenyl **(40)** (91)

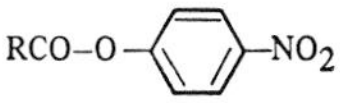

(37) 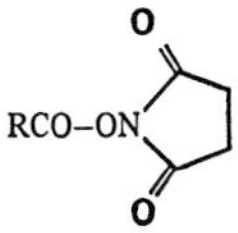 (38) 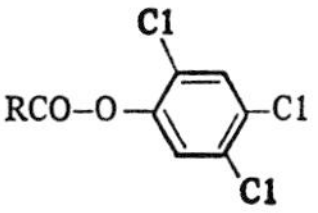 (39) 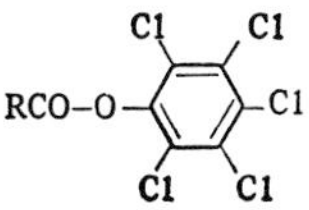 (40)

esters. 8-Hydroxyquinoline esters **(41)** (92) form peptide bonds rapidly through neighboring group assisted catalysis.

Criteria: (i) Racemization, generally not a problem with

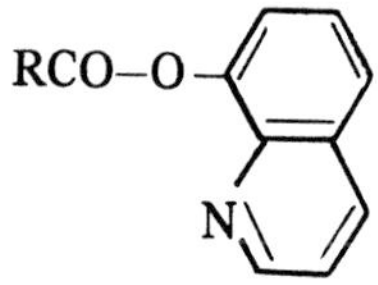

RCO–O– [8-quinolyl]

(41)

active esters of acylamino acids, can become considerable with larger peptides. (ii) Few side reactions have been observed. (iii) Work-up is facilitated when the by-product is water-soluble, as N-hydroxysuccinimide. Complete removal of *p*-nitrophenol is sometimes difficult. (iv) Yields are usually high (60% to 99%). (v) Aminolysis proceeds slower than in mixed anhydride or carbodiimide reactions; and one to four days are required for completion of a coupling.

Large peptides are seldom synthesized by using only one method. It is rather more expedient to rely on all four methods in a selective manner, subject to careful evaluation of sequence-dependent requirements. Table 1-II shows a comparison using the criteria discussed above. The strong and weak points of the methods, relative to each other, appear obvious. This will assist in selecting the best suited method for most cases.

TABLE 1-II.

COMPARATIVE EVALUATION OF THE FOUR EFFICIENT METHODS OF PEPTIDE SYNTHESIS

Criteria	Azide	Mixed Anhydride	Carbodiimide	Active Ester
Minimal Racemization	++++	+	++	+++
Minimal Side Reactions	+	+++	++	++++
Work-up	+++	++++	++	+
Yield	+	+++	++	++++
Time Requirement	+	++++	+++	++

++++ Excellent ◄——► + Satisfactory

Strategy of Peptide Chain Assembly

The choices of protecting groups, methods for peptide bond formation, and procedures for cleavage of protecting groups are regarded as the *tactics* of peptide synthesis. The pattern of assembling all amino acid residues into the desired peptide sequence is referred to as the *strategy.* (5) Two principal strategies can be distinguished. In the *fragment condensation* strategy, amino acids are condensed to small intermediate peptides, so-called fragments. These are coupled to large fragments, eventually giving the final product. With *incremental* or *stepwise chain elongation* (93,94) one amino acid residue is added at a time starting from the C-terminal end. Fragment condensation allows greater flexibility in choice of protecting groups and coupling methods but it is impeded by danger of racemization. For this reason the subdivision of a given sequence into fragments is preferentially designed at glycyl or prolyl bonds which may be coupled without racemization. Both strategies have recently been employed in syntheses of secretin. Figure 1-4a depicts the assembly of the hormone sequence by incremental chain elongation. (18) Figure 1-4b shows how the same molecule was synthesized by fragment condensation. (19)

The extent to which side-chains of trifunctional amino acids have to be protected depends on the peptide and the coupling methods to be used. Frequently, serine, threonine, tyrosine, or histidine can be used with unprotected side chains, as in azide or some active ester condensations. For the synthesis of glucagon (16), however, a strategy of "global protection," i.e. protection of all side-chain functions (Fig. 1-5a) was found to be necessary. It provides maximal safety against undesired side reactions and flexibility in choice of coupling methods. On the other hand, presently used protecting groups can cause larger peptides to become insoluble in organic solvents. Because of this, Hirschmann and collaborators (8,95) employed an absolute minimum of blocking groups in synthesizing ribonuclease S-protein. For example, a single side-chain protection was used in nona-

Val-NH_2

Leu-Val-NH_2

Gly-Leu-Val-NH_2

Gln-Gly-Leu-Val-NH_2

Leu-Gln-Gly-Leu-Val-NH_2

Leu-Leu-Gln-Gly-Leu-Val-NH_2

Arg-Leu-Leu-Gln-Gly-Leu-Val-NH_2

Gln-Arg-Leu-Leu-Gln-Gly-Leu-Val-NH_2

Leu-Gln-Arg-Leu-Leu-Gln-Gly-Leu-Val-NH_2

Arg-Leu-Gln-Arg-Leu-Leu-Gln-Gly-Leu-Val-NH_2

Ala-Arg-Leu-Gln-Arg-Leu-Leu-Gln-Gly-Leu-Val-NH_2

Ser-Ala-Arg-Leu-Gln-Arg-Leu-Leu-Gln-Gly-Leu-Val-NH_2

Asp-Ser-Ala-Arg-Leu-Gln-Arg-Leu-Leu-Gln-Gly-Leu-Val-NH_2

Arg-Asp-Ser-Ala-Arg-Leu-Gln-Arg-Leu-Leu-Gln-Gly-Leu-Val-NH_2

Leu-Arg-Asp-Ser-Ala-Arg-Leu-Gln-Arg-Leu-Leu-Gln-Gly-Leu-Val-NH_2

Arg-Leu-Arg-Asp-Ser-Ala-Arg-Leu-Gln-Arg-Leu-Leu-Gln-Gly-Leu-Val-NH_2

Ser-Arg-Leu-Arg-Asp-Ser-Ala-Arg-Leu-Gln-Arg-Leu-Leu-Gln-Gly-Leu-Val-NH_2

Leu-Ser-Arg-Leu-Arg-Asp-Ser-Ala-Arg-Leu-Gln-Arg-Leu-Leu-Gln-Gly-Leu-Val-NH_2

Glu-Leu-Ser-Arg-Leu-Arg-Asp-Ser-Ala-Arg-Leu-Gln-Arg-Leu-Leu-Gln-Gly-Leu-Val-NH_2

Ser-Glu-Leu-Ser-Arg-Leu-Arg-Asp-Ser-Ala-Arg-Leu-Gln-Arg-Leu-Leu-Gln-Gly-Leu-Val-NH_2

Thr-Ser-Glu-Leu-Ser-Arg-Leu-Arg-Asp-Ser-Ala-Arg-Leu-Gln-Arg-Leu-Leu-Gln-Gly-Leu-Val-NH_2

Phe-Thr-Ser-Glu-Leu-Ser-Arg-Leu-Arg-Asp-Ser-Ala-Arg-Leu-Gln-Arg-Leu-Leu-Gln-Gly-Leu-Val-NH_2

Thr-Phe-Thr-Ser-Glu-Leu-Ser-Arg-Leu-Arg-Asp-Ser-Ala-Arg-Leu-Gln-Arg-Leu-Leu-Gln-Gly-Leu-Val-NH_2

Gly-Thr-Phe-Thr-Ser-Glu-Leu-Ser-Arg-Leu-Arg-Asp-Ser-Ala-Arg-Leu-Gln-Arg-Leu-Leu-Gln-Gly-Leu-Val-NH_2

Asp-Gly-Thr-Phe-Thr-Ser-Glu-Leu-Ser-Arg-Leu-Arg-Asp-Ser-Ala-Arg-Leu-Gln-Arg-Leu-Leu-Gln-Gly-Leu-Val-NH_2

Ser-Asp-Gly-Thr-Phe-Thr-Ser-Glu-Leu-Ser-Arg-Leu-Arg-Asp-Ser-Ala-Arg-Leu-Gln-Arg-Leu-Leu-Gln-Gly-Leu-Val-NH_2

His-Ser-Asp-Gly-Thr-Phe-Thr-Ser-Glu-Leu-Ser-Arg-Leu-Arg-Asp-Ser-Ala-Arg-Leu-Gln-Arg-Leu-Leu-Gln-Gly-Leu-Val-NH_2

1 2 3 4 5 6 7 8 9 10 11 12 13 14 15 16 17 18 19 20 21 22 23 24 25 26 27

Figure 1-4. Strategies of peptide chain assembly. (A) Incremental (stepwise) chain elongation; secretin synthesis.

peptide sequence 77–85 (Fig. 1-5a&b). "Minimal protection" restricts, however, the choice of methods for peptide bond formation.

Rapid methodological progress, as discussed above, has paved the way for impressive achievements in syntheses of

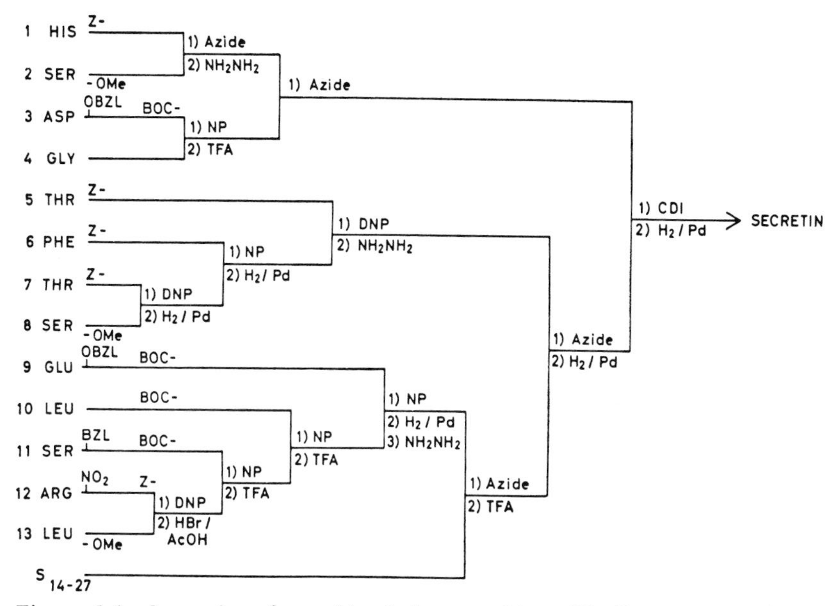

Figure 1-4. Strategies of peptide chain assembly. (B) Fragment condensation; secretin synthesis.

biologically important peptides. A few representative examples are compiled in Table 1-III. A particularly active sector of peptide chemistry has dealt with studies on structure-activity correlations by analog synthesis. To-date, the total number of synthetic analogs of biologically active peptides amounts to well over a thousand. This type of work

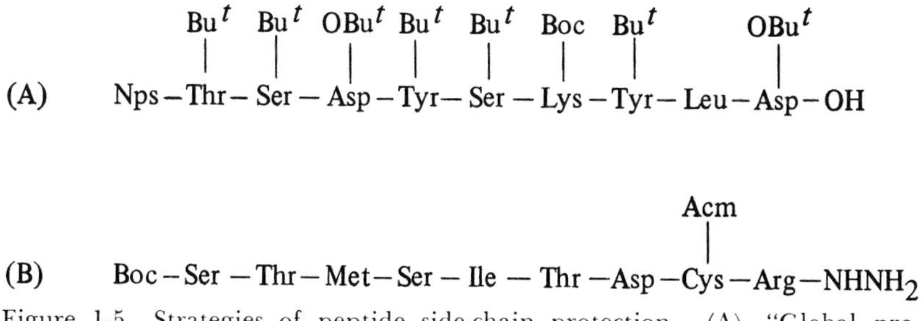

(A) Nps–Thr(Bu^t)–Ser(Bu^t)–Asp(OBu^t)–Tyr(Bu^t)–Ser(Bu^t)–Lys(Boc)–Tyr(Bu^t)–Leu–Asp(OBu^t)–OH

(B) Boc–Ser–Thr–Met–Ser–Ile–Thr–Asp–Cys(Acm)–Arg–$NHNH_2$

Figure 1-5. Strategies of peptide side-chain protection. (A) "Global protection" of a nonapeptide of glucagon sequence 7–15. (B) "Minimal protection" of a nonapeptide of bovine ribonuclease A sequence 77–85. Both nonapeptides contain eight functional side-chain groups.

TABLE 1-III.

Examples of Achievements by Peptide Synthesis in Solution

Peptide and Reference	*Size* [a]
RIBONUCLEASE S-PROTEIN	104
Hirschmann *et al.* 1969 (8)	
ADRENOCORTICOTROPIN	39
Porcine: Schwyzer and Sieber 1963 (96)	
Human: Bajusz *et al.* 1967 (97)	
RIBONUCLEASE T_1- (12-47)	36
Beacham *et al.* 1971 (98)	
Camble *et al.* 1972 (99)	
CALCITONIN	32
Porcine: Rittel *et al.* 1968 (100)	
Guttmann *et al.* 1968 (101)	
Human: Sieber *et al.* 1968 (102)	
Salmon: Guttmann *et al.* 1969 (103)	
INSULIN	30
Meienhofer *et al.* 1963 (13)	21
Katsoyannis *et al.* 1964 (14)	
Kung *et al.* 1965 (15)	
GLUCAGON	29
Wünsch 1967 (16)	
SECRETIN	27
Bodanszky *et al.* 1966 (104)	
GASTRIN	
Sheppard *et al.* 1964 (17)	17

[a] Amino acid residues.

has focussed on the neurohypophyseal hormones oxytocin and vasopressin, (105) on gastrin, (21) on angiotensin (106) and other kinins, (107) on peptide antibiotics, such as gramicidin S (108) or valinomycin, (109) and on ribonuclease S-peptide. (110,111) An interesting and therapeutically useful development has been syntheses of highly potent N-terminal parts of adrenocorticotropin, ACTH. In these analogs the N-terminal serine residue was replaced by D-serine or β-alanine and the C-terminals were derivatized with amide groups. This prevents or slows enzymatic degradation of the hormones after systemic administration and resulted in remarkable increases of biological potency over the 100 IU/mg* of native ACTH. The pentacosapeptide derivative, [D-Ser^1, Nle^4, Val-NH$_2^{25}$]-β-ACTH-**(1–25)** (113) possesses 525IU/mg activity, the octadecapeptide [D-Ser^1, Lys^{17}, Lys-NH$_2^{18}$]-β-ACTH-**(1–18)** (114) has a po-

* The ascorbic acid depletion assay (112) was used for comparison.

tency of 800 IU/mg, and [β-Ala[1]]-ACTH-**(1–17)**-heptadecapeptide-4-amino-*n*-butyl-amide (115) exhibits even slightly more than 800 IU/mg activity.

Scope and Limitation

A protein containing 104 amino acid residues, ribonuclease S-protein, has been synthesized. Amounts, however, were minute and repetition on a larger scale is to be awaited. Most other syntheses were concerned with peptides having less than 40 constituent amino acid residues which can be prepared satisfactorily with modern methodology. Presumably, synthesis of peptides containing 60 to 80 amino acids in useful amounts should become feasible in the near future. Promising new protecting groups have been developed, but even more are needed, along with a larger variety of procedures for the selective removal of protecting groups.

The main limitations of peptide synthesis in solution are: (a) insolubility of large protected peptides in organic solvents, (b) low reaction rates, (c) increased danger of racemization with increasing product-size, (d) mounting difficulties in purifying and analyzing increasingly larger peptides, and (e) very large work load.

It is hoped that future methodological developments will further increase the size of peptides that may be prepared by solution synthesis. To achieve this, the following are required. Fragment coupling must be accelerated, perhaps through catalysis. More powerful solvents and more protecting groups must provide increased solubility and cleavage under more selective conditions. Finally, racemization must be more rigidly suppressed or controlled.

SOLID-PHASE PEPTIDE SYNTHESIS

Principle and Features

Merrifield's method is based on a phase separation. The growing peptide is covalently attached to a solid support throughout a synthesis. Coupling reagents, by-products, and side-products are kept in solution. They are rapidly separable by simple filtration from the main product which always remains in the same reaction vessel. A few rapid successive washing operations thus replace the time consuming work-up and, frequently difficult, purification of peptides in conventional synthesis.

The standard solid-phase procedure is schematically depicted in Figure 1-6. Protecting groups and reagents of proven usefulness in conventional peptide synthesis are em-

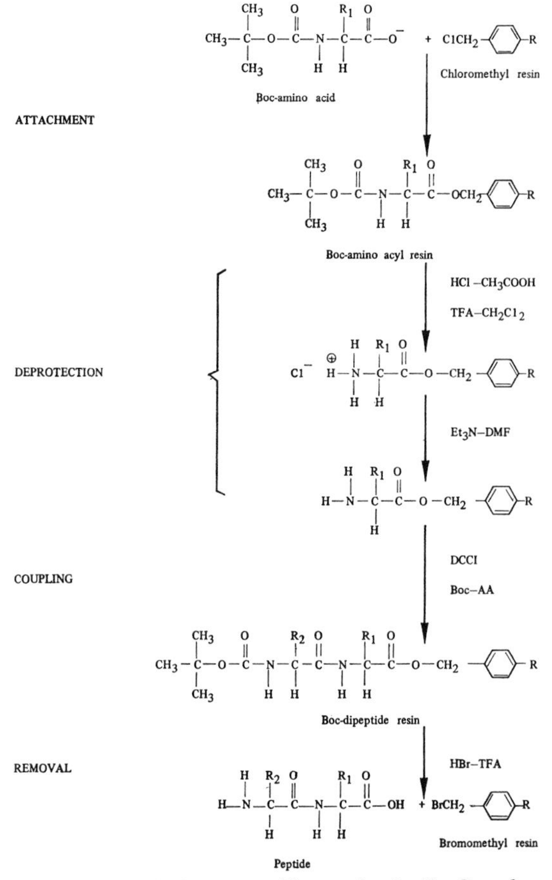

Figure 1-6. Scheme of solid-phase peptide synthesis R, Copolystyrene-divinylbenzene; TFA, trifluoracetate acid; Et_3N, triethylamine; DMF, dimethylformamide; DCCI, dicyclohexylcarbodiimide; Boc-AA, Nα-*tert*-butyloxycarbonylamino acid.

ployed. The most useful solid support is beaded polystyrene crosslinked by 2 percent, or, preferably, 1 percent divinylbenzene. After chloromethylation the C-terminal amino acid is attached to it through a benzyl ester bond ("attachment" stage in Fig. 1-6). The *tert*-butyloxycarbonyl group **(5)** (31) is most frequently used for the protection of the α-amino function. This Nα-protecting group is cleaved with hydrogen chloride in acetic acid or with anhydrous trifluoroacetic acid, alone (116–118) or in methylene chloride. (6) The amino group, which is left in protonated form, is then converted to free amine by treatment with triethylamine in methylene chloride. The deprotection step thus requires two successive operations, (see Fig. 1-6). Coupling with the next *tert*-butyloxycarbonylamino acid is carried out in methylene chloride with the aid of dicyclohexylcarbodiimide. (74) The reaction requires normally one to six hours using four-fold molar excesses of reactants. After each reaction the solid is washed several times with organic solvents to insure complete removal of all reagents, excess *tert*-butyloxycarbonyl amino acids and byproducts. The peptide chain is elongated by alternating successive repetitions of deprotection and coupling steps, called "cycles." After completing the required number of cycles, the peptide is cleaved from the resin by treatment with hydrogen bromide in anhydrous trifluoroacetic acid. (119) Alternatively, anhydrous hydrogen fluoride in the presence of anisole (120) is used. For the great majority of solid-phase syntheses this "standard" Merrifield procedure, or similar procedures with minor variations, have been used. Table 1-IV contains a schedule for experimental performance of one cycle according to the standard procedure, which proved to be particularly satisfactory for syntheses of bradykinin **(42)** and angiotensin **(43)** peptides.

Arg-Pro-Pro-Gly-Phe-Ser-Pro-Phe-Arg **(42)**
Bradykinin

Asp-Arg-Val-Tyr-Ile-His-Pro-Phe **(43)**
5-Isoleucine-Angiotensin II

TABLE 1-IV.

The "Standard" Schedule of Solid-Phase Peptide Synthesis (121)

Operation No.	*Reagent, Solvent* [a]	*Time*
1	Glacial acetic acid wash (3 times) [b]	
2	1 *N* HCl in glacial acetic acid[c]	30 min
3	Glacial acetic acid wash (3 times)	
4	Ethanol wash (3 times)	
5	Dimethylformamide wash (3 times)	
6	Triethylamine in dimethylformamide (1:10)	10 min
7	Dimethylformamide wash (3 times)	
8	Methylene chloride wash (3 times)	
9	Boc-amino acid (4 equiv) in methylene chloride	10 min
10	Dicyclohexylcarbodiimide (4 equiv) in methylene chloride	2 hr
11	Methylene chloride wash (3 times)	
12	Ethanol wash (3 times)	

[a] For 1 g of resin 6 ml of solvent was used.
[b] A wash operation involves 3 minutes mixing followed by filtration.
[c] Alternatively, 4 *N* HCl in dioxane has been used (122).

Solvents and reagents are introduced into a suitable reaction vessel in a timed sequence. Similarly the removal of solutions from the vessel containing the peptide resin is attained by timed application of suction or pressure. Figure 1-7 shows Merrifield's design of the reaction vessel and a simple manually controlled shaking device. All reactions throughout a synthesis are conducted in the vessel without transfer of solid material. Solvents and reagent solutions are added through a side arm. Mixing is by slow, gentle rocking. After reactions, the liquid phase is removed by suction through the fritted glass disk. The resin containing the covalently bound peptide is retained inside the vessel and is ready for the next reagent or solvent to be added. Up to 50 gm and as little as a few milligrams of resin have been handled using different sizes of this type of vessel. Modified designs feature improved mixing, temperature control, larger operating capacity, or analytical process control. (123–126)

The principal motivation of Merrifield in designing the solid-phase method was an automation of peptide synthesis. (127,128) A mechanized programmed apparatus was planned and built. It features two main components, a control system and a reactor system. (129) A stepping drum

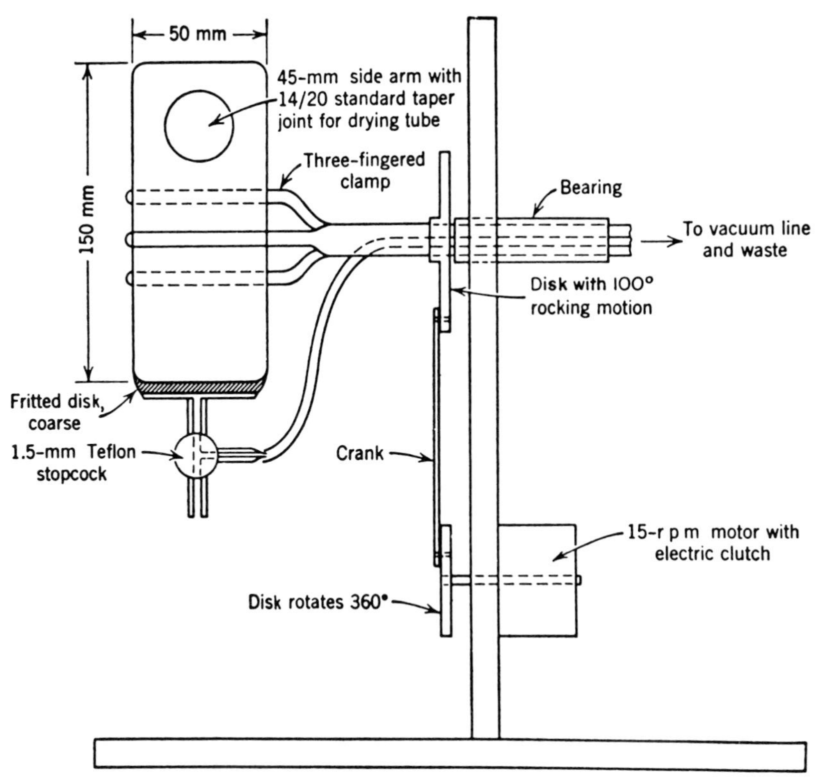

Figure 1-7. A manually operated vessel and shaker for solid-phase peptide synthesis. From Merrifield.[26] Reproduced with permission of Academic Press, Inc.

programmer controls the operations of the vessel shaker, solution pump, distributor valves, air vents, vacuum valves, and timers. The reactor system comprises the vessel, the distributor valves for solvents and reagents, and storage bottles for these liquids, as shown schematically in Figure 1-8. This instrument further reduces the work load of peptide synthesis. However it requires an operator possessing experience in both mechanics and electronics.

In brief, four distinct features mark the solid-phase method. (1) Dramatic time saving due to speed and simplicity of operation is the most important asset. One to six amino acid residues per day may be incorporated into growing peptide

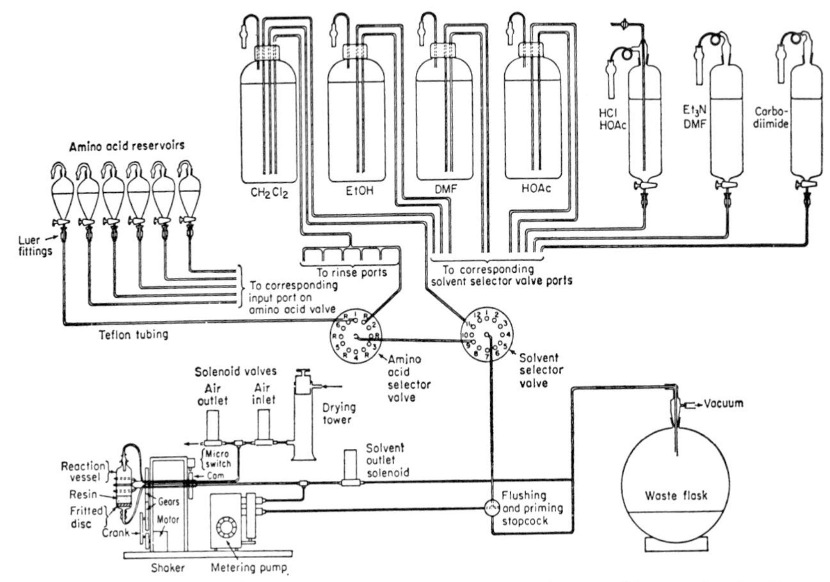

Figure 1-8. Scheme of an apparatus for "automated" peptide synthesis. From Merrifield *et al.*[129] Reproduced with permission of American Chemical Society, Inc.

chains. By comparison, syntheses of larger peptides by conventional methods in solution may require 50 times or more man hours. (2) Insolubility problems interfering with syntheses of larger peptides in solution are eliminated. (3) The procedure is easy and convenient to perform. (4) The procedure has been mechanized and automated.

The method has quickly found wide and frequent application. A large number of biologically active peptides and analogs have been synthesized. Some of these achievements are listed in Table 1-V. Further details and more complete listings may be found in several reviews. (26,27,145,146) Chemical and methodological aspects have been authoritatively discussed by Merrifield. (26) Practical experimental instruction can be found in a recent book. (147)

Problems and Shortcomings

Criticism of solid-phase peptide synthesis has been directed mainly at product heterogeneity. As with other se-

TABLE 1-V.

Examples of Synthetic Achievements by the Merrifield Solid-Phase Method

Peptide and Reference	*Size* [a]
HUMAN GROWTH HORMONE Li and Yamashiro 1970 (9)	188 [b]
RIBONUCLEASE A Gutte and Merrifield 1969 (6,7)	124
CYTOCHROME C Sano and Kurihara 1969 (130)	104
RIBONUCLEASE T_1 Izumiya *et al.* 1971 (131)	104
ACYL CARRIER PROTEIN- (1-74) Hancock *et al.* 1971 (132)	74
BOVINE BASIC TRYPSIN INHIBITOR Noda *et al.* 1971 (133)	58
FERREDOXIN Bayer *et al.* 1968 (134)	55
STAPHYLOCOCCAL NUCLEASE- (6-47) Ontjes and Anfinsen 1969 (135)	42
PARATHYROID HORMONE- (1-34) Potts *et al.* 1971 (136)	34
INSULIN Marglin and Merrifield 1966 (137)	30 21
ADRENOCORTICOTROPIN- (1-19) Blake *et al.* 1972 (138)	19
α-MELANOTROPIN Blake and Li 1971 (139)	13
ANTAMANIDE Wieland *et al.* 1969 (140)	10
GRAMICIDIN S Ohno *et al.* 1971 (141)	10

[a] Amino acid residues.
[b] According to the originally proposed structure (142,143). For a revised complete structure see Li and Dixon (144).

quential procedures, the method depends decisively upon 100 percent completeness of all reactions. This is not the case at the present state of development where average yields range from 97 percent to 99 percent per peptide coupling. As shown in Table 1-VI, these yields are inadequate for syntheses of larger peptides or proteins. With 98 percent constant coupling yield per step, for example, the overall yield of ribonuclease A would be only 8.3 percent. The other 91.7 percent of product would be closely related contaminants, consisting of over 4 billion different species of failure sequences (147a) or deletion peptides,* truncated peptides,

* The number of different molecules is derived from the binomial distribution, if failure sequences are formed in each cycle. Size range of these deletion peptides is from 118 to 123 constituent amino acid residues.

TABLE 1-VI.

Overall Yields of Peptides Prepared *via* Incremental Chain Elongation as in Solid-Phase Synthesis; Calculated [a] for Several Constant Yields per Step

Protein	*Amino Acid Residues*	*Overall Yield of Target Peptide*			
		Yield per Step			
		95%	*98%*	*99%*	*99.9%*
Human Growth Hormone	190	0.006	2.2	15.0	82.8
Ribonuclease A	124	0.2	8.3	29.1	88.4
Cytochrome c	104	0.5	12.4	35.5	90.2
Bovine Trypsin Inhibitor	58	5.4	31.6	56.4	94.5
Ferredoxin	55	6.3	33.6	58.1	94.7
Insulin B-Chain	30	22.6	55.7	74.7	97.1
Insulin A-Chain	21	35.8	66.8	81.8	98.0

[a] Based on the starting (C-terminal) amino acid.

and diastereoisomers originating from racemization. Complex "microheterogeneous mixtures," such as this, pose most formidable problems of product purification. Even the most efficent multistage fractionation procedures are insufficient. At best, a partial purification can be achieved. At present, solid-phase synthesis might provide close to homogeneous peptides* up to pentadeca- or eicosapeptide (20 amino acid) size. Synthesis of acceptably pure proteins will require improvements that guarantee 99.9 percent yields per cycle. To achieve this, both the deprotection stage and the coupling stage (see Fig. 1-6) will have to be optimized. Obviously, the principal feature of the method, attachment of peptide to solid throughout a synthesis, becomes also the biggest disadvantage when reactions do not go to 100 percent completion. It prevents intermediate purification. As a consequence unreacted components are retained and complex mixtures accumulate.

Another problem is loss of peptide from solid-phase during synthesis. This has first been reported for apoferredoxin synthesis (134) where titratable free amine values dropped to 38 percent after 54 cycles (Fig. 1-9). In the synthesis of ribonuclease A, only 17 percent of the peptide was retained after 123 cycles. (7)

Incomplete stability of side-chain protecting groups, such as Nω-benzyloxycarbonyl, has resulted in formation of

* After extensive product purification.

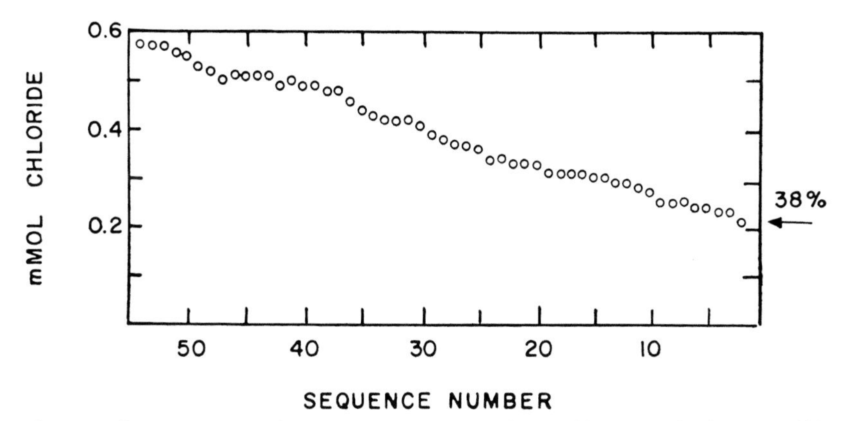

Figure 1-9. Decreasing free amine content of peptide-resin during a solid-phase synthesis of apoferredoxin over 54 cycles is shown from triethylamine hydrochloride titration values after each deprotection step. From Bayer *et al.*[134] Reproduced with permission of Pergamon Press.

branched peptide chains in subsequent cycles. Product mixtures were contaminated with peptides containing more amino acid residues than expected (Fig. 1-10). (148) Product decomposition during removal of peptide from solid support at the end of synthesis has often been a further complication. (7,9) Another shortcoming continues to be the lack of highly sensitive analytical controls for detection of incomplete solid-phase reactions, although a few useful colorimetric and titration tests were developed. Some efforts at improving these deficiencies will be described below.

New Methodological Developments

Promising developments are resin modifications, surface supports, improved protecting group combinations, and effective analytical controls. Functional cross-linked polystyrene derivatives, forming covalent bonds other than benzyl esters with peptides, have been studied for three purposes (1) Linkages are sought that would provide mild cleavage of completed peptides without product decomposition. A promising example is *tert*-alkyl alcohol resin **(44)**. (149) (2) Increased stability throughout a synthesis, offered by acylsulfonamide bonds **(45)** (150) should reduce loss of

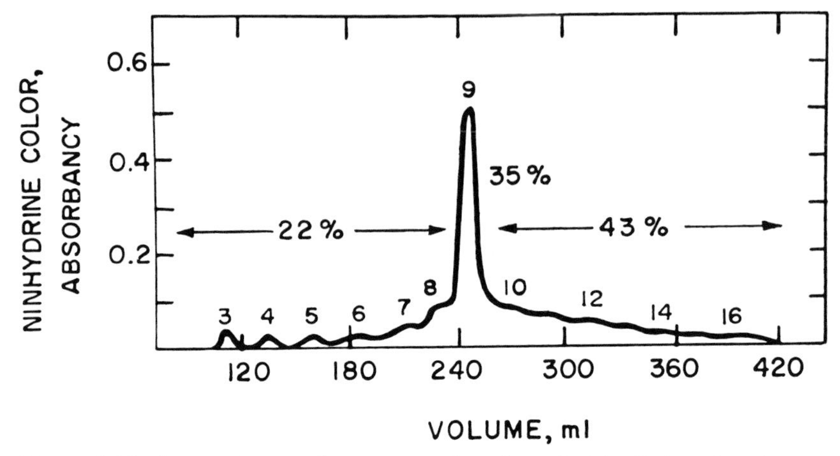

Figure 1-10. Ion exchange chromatography of crude Nα-dinitrophenyl nonalysine, prepared by solid-phase synthesis. Nα-Boc-Nε-Z-lysine was used. Nα-Deprotection was effected by 1*N* HCl in acetic acid. Numbers indicate lysine residues in products. Components containing over 9 lysine residues originated from side-chain branching. From Yaron and Schlossman.[148] Reproduced with permission of American Chemical Society, Inc.

$$HO-C(CH_3)_2-CH_2CH_2-C_6H_4-Resin$$

(44)

$$RCO-NHSO_2-C_6H_4-Resin$$

(45)

peptide from solid support. (3) Preparation of protected peptides suitable for further fragment condensation is provided by a resin possessing *tert*-alkyloxycarbonylhydrazide function **(46)**. (151)

$$H_2NNH-COO-C(CH_3)_2-CH_2CH_2-C_6H_4-Resin$$

(46)

Surface supports are expected to provide faster and more complete reactions by elimination of diffusion barriers. Such solids should allow column operation because they do not swell and shrink. Porous glass functionalized with hydroxylbenzyl groups (152,153) and polystyrene grafted onto Teflon® or polytrifluorochloroethylene (154) have been developed.

In our own work, attachment of peptide to resin through functional groups in amino acid side-chains has been proposed. (155) Lysine-vasopressin has been synthesized after attachment of peptide to chloroformoxymethylated resin through the lysine ε-amino group (Fig. 1-11). Side-chain attachment through other amino acid residues offers possibilities as disulfide bonds (cysteine), amide linkages (asparagine, glutamine), or ether bonds (serine, threonine, tyrosine).

Improved protecting group combinations require either (1) N^{α}-protection of an acid sensitivity higher than that of the *tert*-butyloxycarbonyl group, or (2) side-chain protection of higher acid-stability than that of benzyl derivatives. The 2-(*p*-biphenylyl)-isopropyloxycarbonyl **(47)** (156,157) and the 2-benzoyl-1-methylvinyl **(48)** (158,159) groups are

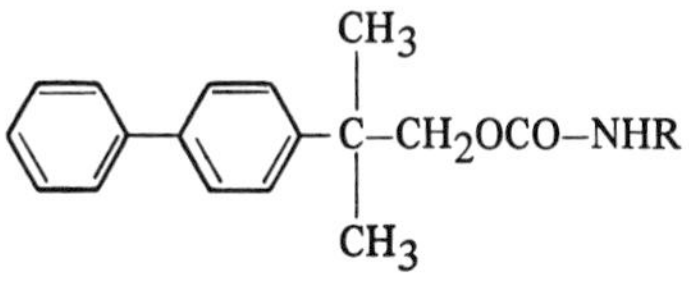

CH_3 / C–CH_2OCO–NHR / CH_3

(47)

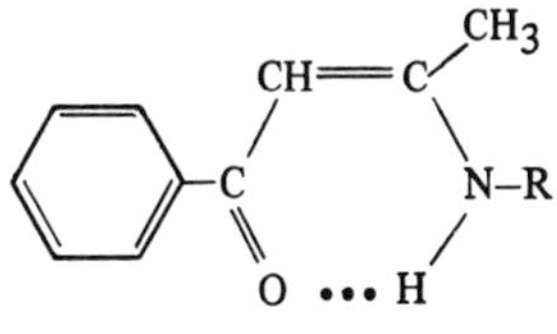

CH=C–CH_3 / C N–R / O ••• H

(48)

cleaved by very mild acid treatment (1% trifluoroacetic acid in methylene chloride) and should be advantageous when used in combination with side-chain protection by benzyl derived groups. A recent synthesis of gramicidin S in which the p-nitrobenzyloxycarbonyl group (160) was used for ω-amino protection is an example of the second approach.

Sensitive and rapid analytical controls are needed for solid-phase synthesis because elemental analysis is not applicable

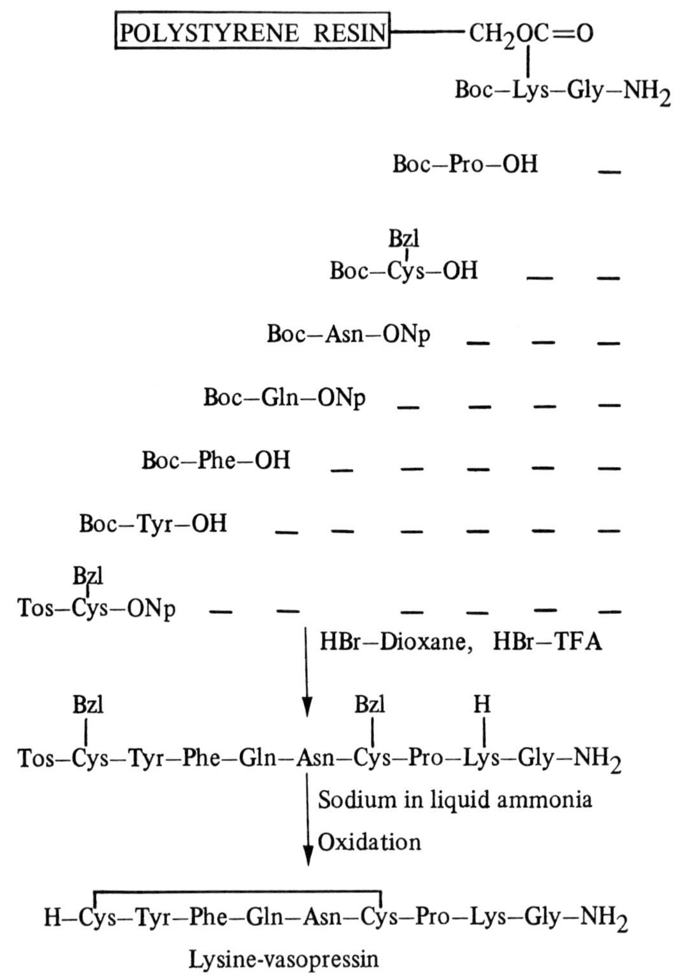

Figure 1-11. Peptide to resin attachment through an amino acid side-chain; synthesis of lysine-vasopressin.[155]

and amino acid analysis after each cycle (122) is too slow. In the "Dorman titration" resin amine content is determined by chloride titration after incubation of peptide-resin with pyridine hydrochloride. (161) The colorimetric "Kaiser test" (162) is presently the most rapid control procedure and very reliable at a sensitivity of approximately $\pm$ 1 percent. The ninhydrin reaction is used to estimate semiquantitatively

TABLE 1-VII.

Estimation of Residual Amine on Peptide-Resin by Ninhydrin Color
[From Kaiser *et al.* (162)]

% *Residual Amine* [a]	*Ninhydrin Color Test* [b]
24	Beads dark blue—dark blue solution
16	Beads dark blue—moderately blue solution
6	Beads moderately blue—light blue solution
0.6	Beads light blue—light greenish solution
0	Beads white—yellow solution

[a] Based on a total peptide content 0.52 mmol/g resin.
[b] With N-terminal prolyl or β-benzyl aspartyl residues brown to reddish-brown colors are observed.

amounts of free amine from color intensities and shades, Table 1-VII, and evaluate the completeness of the coupling step. The test is done on 10 mg resin samples and requires only about 10 minutes for completion. No tests, however, are yet precise enough. A 0.1 percent sensitivity will be required if 99.9 percent yield per cycle is to be attained.

Scope and Limitations

The solid-phase method was designed for the chemical synthesis of proteins. The efficacy has been tested in part. Nuclease and growth hormone syntheses indicated that coupling efficiency, i.e. rate and yield of amino acid incorporation, remained satisfactory throughout 123 and 187 cycles, respectively. The method thus functions for syntheses of medium sized proteins. On the other hand, present coupling yields of 95 to 99 percent per cycle lead to complex product mixtures. Obviously solid-phase peptide synthesis must be followed by extensive product purification. The size limit for product homogeneity appears to be at pentadecapeptides. Under favorable conditions homogeneous eicosapeptides might be obtained after extensive purification. Not only are improvements in synthetic procedure required to increase yields beyond 99 percent per step, but also more powerful fractionation methods are needed to obtain uniform peptides containing more than 20 amino acid residues.

Some biological problems can be studied by using mixtures obtained by solid-phase synthesis. The method does not appear to suffer from an upper size restriction: *protein mixtures can readily be prepared.* However, each investigator must ponder and weigh carefully two critical questions. (1) How useful are these mixtures for the particular study of biological, biochemical, or medicinal problems. (2) Which correlation of synthetic mixtures to natural products can be safely deduced.

In sum, to reduce product heterogeneity average coupling efficiency must be increased to 99.9 percent, while feedback control with 0.1 percent sensitivity is required for full automation. In addition, new solid supports for mild removal conditions, improved combinations of protecting groups, ultrasensitive analytical tools and more effective fractionation procedures will also be needed.

COMPARISON AND CHOICE OF METHOD

Comparison of Solid-Phase and Solution Methods of Peptide Synthesis

To assist the difficult choice between conventional solution or solid-phase methods for synthesis of a given peptide comparative evaluation of both approaches will be made focussing on strong and weak points of each. By tabulating, a cross-correlation point by point can be made (Table 1- VIII). Solid-phase synthesis is very fast and convenient, while solution synthesis is slow and laborious. Solution synthesis of larger peptides is impeded by low solubility, failure or low yield in fragment condensation, and danger of considerable racemization. These difficulties do not exist in solid-phase synthesis. In fact, elimination of the very serious problem of low solubility of larger peptides in organic solvents is one of the most important experimental innovations of the solid-phase method. No special organic chemistry experience is required to carry out solid-phase synthesis, while the requirement of great expertise has been one of the very real restrictions to conventional peptide synthesis.

Solid-phase syntheses, on the other hand, usually produce microheterogeneous mixtures of peptides with more than

TABLE 1-VIII.

GENERAL COMPARISON BETWEEN SOLID-PHASE AND SOLUTION SYNTHESIS OF PEPTIDES

Solid-Phase Synthesis		*Conventional Synthesis in Solution*
SPEED	STRONG POINTS	PRODUCT : HOMOGENEOUS
NO SOLUBILITY PROBLEM		ADEQUATE ANALYSIS
GENERAL SYNTHETIC EXPERIENCE SUFFICIENT		
AUTOMATION POSSIBLE		PURIFICATION OF INTERMEDIATES
PRODUCT : MIXTURE a) < 100% Cleavage of N-protection b) < 100% Peptide Bond Formation c) Small Amount of Racemization (0.1%)	WEAK POINTS	SLOW, LABORIOUS INSOLUBILITY
REMOVAL from SOLID PHASE a) Incomplete b) Product Degradation		FAILURE of FRAGMENT CONDENSATION RACEMIZATION DANGER
EXTENSIVE PRODUCT PURIFICATION NEEDED		MUCH SPECIAL EXPERIENCE REQUIRED
NO ADEQUATE ANALYSIS (0.1 to 0.5%)		

10 to 20 constituent amino acids. Available analytical methods can neither define them nor assess product homogeneity after purification. In contrast, solution synthesis, generally affords homogeneous peptides because intermediates are purified after each step. Analytical procedures are adequate because they also are applied after each condensation. In solution synthesis reaction mixtures usually consist of only a *few* predictable components.

In brief, the strength of solid-phase synthesis lies in *operational efficiency* as compared to *product quality* which is the strength of the conventional solution methods. (27) When identical equally pure peptides are obtained by either approach, the solid-phase method is preferred by virtue of its time saving, easy manipulation, and higher yields.

The following examples are direct comparisons of syntheses by both approaches in the same laboratory in which the solid-phase synthesis was superior: antamanid **(49)**, (140,163) the antidote to the fungal toxin am-

```
┌─Pro ──── Phe ──── Phe ──── Val ──── Pro ─┐
│                                          │   (49)
└─Pro ←─── Phe ←─── Phe ←─── Ala ←─── Pro ←┘

┌─Val ──── Orn ──── Leu ──── D–Phe ──── Pro ─┐
│                                            │   (50)
└─Pro ←─── D–Phe ←─── Leu ←─── Orn ←─── Val ←┘

                      NO2·Z                              NO2·Z
                        |                                  |
Meoz – D–Phe–Pro–Val–Orn–Leu–D–Phe–Pro–Val–Orn–Leu–NHNH2          (51)
```

anitin, (164) and gramicidin S **(50)**. (141,165) The latter gave a crystalline protected linear-chain intermediate **(51)** which was obtained by resin hydrazinolysis (166,167) in 97 percent overall yield.*

Conversely, in another comparative study, solid phase synthesis was judged inadequate in our laboratory for the preparation of hepta-γ-L-glutamic acid when using benzyl ester protection (Boc-Glu-OBzl). (168) The product mixture could not be completely fractionated by several column chromatographic procedures and hepta-γ-glutamic acid could not be obtained in homogeneous form. Conventional solution synthesis using Z-Glu-OBu^t and mixed anhydride (80) condensation afforded all protected intermediate oligomers in crystalline, analytically pure form. Considerably less time was spent than in the futile attempts to purify the mixtures obtained by solid-phase synthesis.

Project-Oriented Choice of Method

Even with clear understanding of comparative scopes and limitations of both solid-phase and solution methods it will often be difficult to decide which should be used for a given synthesis. Other relevant considerations include personal experience in the field and available time, manpower, and resources. Decisive considerations in choosing between the

* The yield is based on the starting C-terminal resin-bound amino acid (leucine).

TABLE 1-IX

Criteria for Strategic Use and Project Oriented Choice

	Size (Amino Acid Residues)					
Purpose	3–10	10–20	20–50	50–100	Small Protein	Large Protein
Intrinsic Activities	SPS*	SPS	SPS	SPS	SPS	SPS
Therapeutic Use	SPS	SPS	SPS	SPS	SPS	SPS
Active Core	SPS	SPS	SPS	SPS	SPS	....
Identify Active Principle	SPS	SPS,SOL†	SOL	SOL	....	....
Proof of Structure	SPS,SOL	SOL	SOL	....	....	....
Conformation Study	SPS,SOL	SOL	SOL	....	....	....
Minimal Active Size	SOL	SOL	SOL	....	....	....
Analogs						
I) Well-Known Biology						
a) Highly Active	SPS	SPS	SPS	SPS	SPS	SPS
b) Weakly Active	(SPS),SOL	SOL	SOL	....	....	....
II) Unknown Biology	SOL	SOL	....	....	....	....

* SPS, Solid-phase peptide synthesis.

† SOL, Conventional peptide synthesis in solution.

two approaches are the nature of the project under study and the characteristics of the peptide. One important criterion would obviously be the peptide size. If the purpose of the work is correlated with the size of peptide a diagram can be constructed, Table 1-IX.

Peptides with over 100 amino acid residues can, at this time, only be prepared by solid-phase synthesis. Intrinsic biological activity can ideally be explored *via* this method.* Production for therapeutic use, preparation of an active core of a larger protein, or recognition of an unknown active center by rapid synthesis of many fragments can all be carried out by the solid-phase method.

When the homogeneity of the peptide is of prime importance, however, applicability of the solid-phase method becomes limited to smaller size peptides. Here one must be reasonably sure that single molecular entities and not mixtures are obtained. These peptides are for the exploration of specific problems such as identification of an unknown active principle, structural proof, or conformational studies. To

* For example, the origin of some lactogenic activity present in isolated natural human growth hormone was shown by a solid-phase synthesis (9) to be intrinsic rather than due to possible contamination by lactogenic hormone.

determine the smallest active fragment of a peptide, one definitely does not want to have a contaminating failure peptide in the synthetic product which, by chance, might happen to possess activity. Here only solution synthesis will suffice.

For synthesis of highly potent analogs of biologically active peptides the solid-phase method can safely be employed. In fact, it is ideally suited for readily providing large numbers of analogs required for the exploration of structure-activity relationships. When activities drop below the 5 to 10 percent level, however, solid-phase procedures become inadequate. An activity of 1 percent could, for example, be due to 1 percent of a contaminating failure peptide possessing full activity, or it could be due to 10 percent of failure peptides having about 10 percent activity.

Criteria other than size of peptide to be prepared should certainly be considered as well when choosing between solid-phase and solution methods and these might be plotted similarly. The essential point which I wish to emphasize is that some peptide syntheses can be pursued with advantage by solid-phase synthesis and others must be carried out by classical synthesis, depending upon the nature of the desired information.

CONCLUDING REMARKS

In this review I have attempted to delineate the present state of the art of chemical synthesis of peptides and to show that the era of chemical synthesis of proteins has been entered. Experimental shortcomings and problems have been outlined and the need for substantial methodological improvements has been demonstrated. The capacity of modern peptide chemistry for preparation of many biologically active peptides has been documented. Only a few examples of the multitude of synthetic achievements could be mentioned within the limits of this chapter.

The potential of chemical peptide and protein synthesis to contribute to nutrition research may not be obvious. It deserves some concluding comments. A chance discovery and a prediction of future requirements might serve as examples.

In the course of synthetic studies on C-terminal gastrin peptides (169) L-aspartyl-L-phenylalanine methyl ester **(52)** was, quite incidentally, found to have an extremely sweet taste. Further studies showed that this dipeptide ester is 100–200 times sweeter than sucrose. Moreover, the peptide

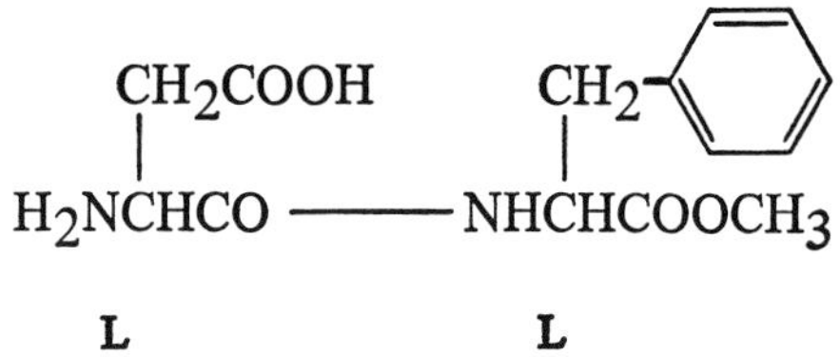

$$\underset{\mathbf{L}}{H_2N\overset{\overset{\displaystyle CH_2COOH}{|}}{C}HCO} \text{———} \underset{\mathbf{L}}{NH\overset{\overset{\displaystyle CH_2\text{-}C_6H_5}{|}}{C}HCOOCH_3} \qquad (52)$$

ester has a good flavor and no unpleasant aftertaste. Many analogs have been synthesized. Only a few were found to be similarly sweet, such as H-Asp-Phe-OEt and H-Asp-Met-OMe. The implications and potential of these compounds for persons afflicted with diabetes mellitus are obvious.

Due to the high rate of increase of diabetes, the supply of insulin (structure, see Fig. 1-2) from natural sources probably will no longer be adequate within a decade or two. Little may be done to increase insulin supply from cattle. Chemical synthesis may then be required to provide the additional amounts of insulin. Several synthetic approaches were developed some years ago. (13–15) A- and B-chains might be prepared through these or other pathways. Chain combination, however, through the formation of three correctly-paired disulfide bonds, the final stage of insulin synthesis remains problematic. Yields after random co-oxidation of mixtures of reduced chains are usually very low (5 to 30 percent). The exciting discovery of proinsulin (170,171) and elucidation of structure (172) (Fig. 1-12) is an advance of the greatest importance. Proinsulin is a single-chain molecule. Oxidative disulfide bond formation of the reduced hormonogen is an intramolecular, sequence-controlled process affording high recoveries of native proinsulin. This should pave the way to a more efficient strategy

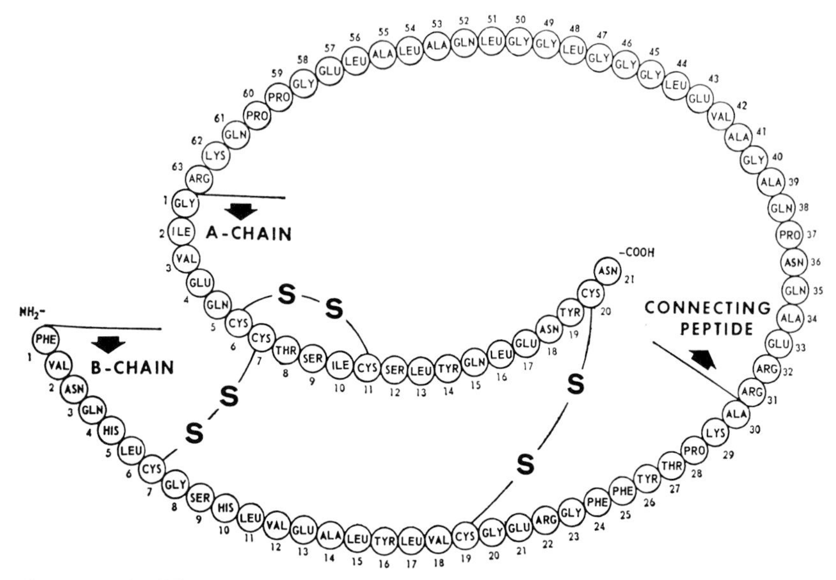

Figure 1-12. The structure of porcine proinsulin. From Chance *et al.*[172] Reproduced with permission of American Association for the Advancement of Science.

of insulin synthesis. Work on total synthesis of proinsulin is already in progress in several laboratories.

Peptide synthesis thus contributes in many ways to nutritional and other areas of medicinal research. Future goals are chemical syntheses of protein molecules, unavailable in nature, for biochemical research. Synthetic products should be obtained in homogeneous form for that purpose. Considerable methodological improvements will be required to achieve this goal. The task of the peptide chemist is to develop necessary techniques.

REFERENCES

1. Abbreviations used for amino acids and the designation of peptides follow the rules of the IUPAC-IUB Commission of Biochemical Nomenclature, in *Biochemistry, 5:* 1445 and 2845, 1966. The following additional abbreviations are used: Acm, S-acetamidomethyl; Boc, *tert*-butyloxycarbonyl; Bu^t, *tert*-butyl; Bzl, benzyl; DCCI, dicyclohexylcarbodiimide; DMF, dimethylformamide; Et_3N, triethylamine; EtOH, ethanol; For, formyl; HOAc, acetic acid; MeOH,

methanol; Meoz, *p*-methoxybenzyloxycarbonyl; NO_2·Z, *p*-nitrobenzyloxycarbonyl; Nps, *o*-nitrophenylsulfenyl; OBu^t, *tert*-butyl ester; OBzl, benzyl ester; OEt, ethyl ester; OMe, methyl ester; ONp, *p*-nitrophenyl ester; Pht, phthalyl; Pyr, pyroglutamic acid (L-pyrrolidone-2-carboxylic acid); Resin, copolystyrene-2% divinylbenzene; TFA, trifluoroacetic acid; Tfa, trifluoroacetyl; Tos, *p*-toluenesulfonyl; TRH, thyrotropin-releasing hormone; Z, benzyloxycarbonyl.—Own work, referred to in this text has been supported by NIH grants C-6516 and FR-05526.

2. Geiger, R.: The synthesis of physiologically active peptides. *Angew Chem, Int Ed, 10:* 152, 1971.
3. du Vigneaud, V., Ressler, C., Swan, J. M., Roberts, C. W., Katsoyannis, P. G., and Gordon, S.: The synthesis of an octapeptide amide with the hormonal activity of oxytocin. *J Am Chem Soc, 75:* 4879, 1953.
4. Schröder, E., and Lübke, K.: *The Peptides.* New York, Academic Press, vols. I and II, 1965 and 1966.
5. Bodanszky, M., and Ondetti, M. A.: *Peptide Synthesis.* New York, Interscience, 1966.
6. Gutte, B., and Merrifield, R. B.: The total synthesis of an enzyme with ribonuclease A activity. *J Am Chem Soc, 91:* 501, 1969.
7. Gutte, B., and Merrifield, R. B.: The synthesis of ribonuclease A. *J Biol Chem, 246:* 1922, 1971.
8. Hirschmann, R., Nutt, R. F., Veber, D. F., Vitali, R. A., Varga, S. L., Jacob, T. A., Holly, F. W., and Denkewalter, R. G.: Studies on the total synthesis of an enzyme. V. The preparation of enzymatically active material. *J Am Chem Soc, 91:* 507, 1969.
9. Li, C. H., and Yamashiro, D.: The synthesis of a protein possessing growth-promoting and lactogenic activities. *J Am Chem Soc, 92:* 7608, 1970.
10. Boler, J., Enzmann, F., Folkers, K., Bowers, C. Y., and Schally, A. V.: The identity of chemical and hormonal properties of the thyrotropin releasing hormone and pyroglutamyl-histidyl-prolinamide. *Biochem Biophys Res Commun, 37:* 705, 1969.
11. Burgus, R., Dunn, T. F., Desiderio, D. M., Ward, D. N., Vale, W., Guillemin, R., Felix, A. M., Gillessen, D., and Studer, R. O.: Biological activity of synthetic polypeptide derivatives related to the structure of hypothalamic TRF. *Endrocrinology, 86:* 573, 1970.
12. Celis, M. E., Taleisnik, S., and Walter, R.: Regulation of formation and proposed structure of the factor inhibiting the release of melanocyte stimulating hormone. *Proc Nat Acad Sci U S, 92:* 5748, 1970.
13. Meienhofer, J., Schnabel, E., Bremer, H., Brinkhoff, O., Zabel, R., Sroka, W., Klostermeyer, H., Brandenburg, D., Okuda, T., and Zahn, H.: Synthese der Insulinketten und ihre Kombination zu insulinaktiven Präparaten. *Z Naturforsch, 18b:* 1120, 1963.

14. Katsoyannis, P. G., Fukuda, K., Tometsko, A., Suzuki, K., and Tilak, M.: Insulin peptides. X. The synthesis of the B-chain of insulin and its combination with natural or synthetic A-chain to generate insulin activity. *J Am Chem Soc, 86:* 930, 1964.
15. Kung, Y.-t., Du, Y.-c., Huang, W.-t., Chen, C.-c., Ke, L.-t., Hu, S.-c., Jiang, R.-q., Chu, S.-q., Niu, C.-i., Hsu, J.-z., Chang, W.-c., Cheng, L.-l., Li, H.-s., Wang, Y., Loh, T.-p., Chi, A.-h., Li, C.-h., Shi, P.-t., Yieh, Y.-h., Tang, K.-l., and Hsing, C.-y.: Total synthesis of crystalline insulin. *Sci Sinica (Peking), 14:* 1710, 1965.
16. Wünsch, E.: Die Totalsynthese des Pankreas-Hormons Glucagon. *Z Naturforsch, 22b:* 1269, 1967.
17. Anderson, J. C., Barton, M. A., Gregory, R. A., Hardy, P. M., Kenner, G. W., MacLeod, J. K., Preston, J., Sheppard, R. C., and Morley, J. S.: Synthesis of gastrin. *Nature, 204:* 933, 1964.
18. Bodanszky, M., Ondetti, M. A., Levine, S. D., and Williams, N. J.: Synthesis of secretin. II. The stepwise approach. *J Am Chem Soc, 89:* 6753, 1967.
19. Ondetti, M. A., Narayanan, V. L., von Saltza, M., Sheehan, J. T., Sabo, E. F., and Bodanszky, M.: The synthesis of secretin. III. The fragment-condensation approach. *J Am Chem Soc, 90:* 4711, 1968.
20. Tracy, H. J., and Gregory, R. A.: Physiological properties of a series of synthetic peptides structurally related to gastrin I. *Nature, 204:* 935, 1964.
21. Morley, J. S.: Structure-function relationships in gastrin-like peptides. *Proc Roy Soc, B 170:* 97, 1968.—Structure-activity relationships. *Fed Proc, 27:* 1314, 1968.
22. Weitzel, G., Weber, U., Hörnle, S., and Schneider, F.: Struktur und Wirkung von Insulin: Synthestische A-Ketten mit variierter Sequenz. In Bricas, E. (Ed.): *Peptides 1968.* Amsterdam, North-Holland, 1968, pp. 222–227.
23. Weitzel, G., Eisele, K., Zollner, H., and Weber, U.: Struktur und Wirkung von Insulin, VII. Verkürzte synthetische B-Ketten. *Hoppe-Seyler's Z Physiol Chem, 350:* 1480, 1969.
24. Weitzel, G., Weber, U., Eisele, K., Zollner, H., and Martin, J.: Struktur und Wirkung von Insulin, VIII. Austausch von Histidin gegen Alanin in synthetischen B-Ketten. *Hoppe-Seyler's Z Physiol Chem, 351:* 263, 1970.
25. Merrifield, R. B.: Solid phase peptide synthesis. I. The synthesis of a tetrapeptide. *J Am Chem Soc, 85:* 2149, 1963.
26. Merrifield, R. B.: Solid-phase peptide synthesis. *Advan Enzymol, 32:* 221, 1969.
27. Meienhofer, J.: Peptide synthesis: A review of the solid-phase method. In Li, C. H. (Ed.): *Hormonal Proteins and Peptides,* Vol. 2. New York. Academic Press, 1973.

28. Meienhofer, J.: Synthesen biologisch wirksamer Peptide. *Chimia, 16:* 385, 1962.
29. Denkewalter, R. G., Schwam, H., Strachan, R. G., Beesley, T. E., Veber, D. F., Schoenewaldt, E. F., Barkemeyer, H., Paleveda, Jr., W. J., Jacob, T. A., and Hirschmann, R.: The controlled synthesis of peptides in aqueous medium. I. The use of α-amino acid N-carboxyanhydrides. *J Am Chem Soc, 88:* 3163, 1966.
30. Bergmann, M., and Zervas, L.: Uber ein allgemeines Verfahren der Peptid-Synthese. *Ber Dtsch Chem Ges, 65B:* 1192, 1932.
31. Anderson, G. W., and McGregor, A. C.: *t*-Butyloxycarbonylamino acids and their use in peptide synthesis. *J Am Chem Soc, 79:* 6180, 1957.
32. Zervas, L., Borovas, D., and Gazis, E.: New methods in peptide synthesis. I. Tritylsulfenyl and *o*-nitrophenylsulfenyl groups as N-protecting groups. *J Am Chem Soc, 85:* 3660, 1963.
33. Zervas, L., and Hamalidis, C.: New methods in peptide synthesis. II. Further examples of the use of the *o*-nitrophenylsulfenyl group for the protection of amino groups. *J Am Chem Soc, 87:* 99, 1965.
34. Schönheimer, R.: Ein Beitrag zur Bereitung von Peptiden. *Hoppe-Seyler's Z Physiol Chem, 154:* 203, 1926.
35. du Vigneaud, V., and Behrens, O. K.: A method for protecting the imidazole ring of histidine during certain reactions and its application to the preparation of L-amino-N-methylhistidine. *J Biol Chem, 117:* 27, 1937.
36. Hillmann, A., and Hillmann, G.: Uber leicht abspaltbare Acylreste bei der Peptidsynthese. *Z Naturforsch, 6b:* 340, 1951.
37. du Vigneaud, V., Dorfmann, R., and Loring, H. S.: A comparison of the growth-promoting properties of D- and L-cystine. *J Biol Chem, 98:* 577, 1932.
38. Fruton, J. S., and Clarke, H. T.: Chemical reactivity of cystine and its derivatives. *J Biol Chem, 106:* 667, 1934.
39. Sheehan, J. C., and Yang, D. D. H.: The use of N-formylamino acids in peptide synthesis. *J Am Chem Soc, 80:* 1154, 1958.
40. Weygand, F., and Csendes, E.: N-trifluoracetyl-aminosäuren. *Angew Chem, 64:* 136, 1952.
41. Kidd, D. A. A., and King, F. E.: Preparation of phthalyl-L-glutamic acid. *Nature, 162:* 776, 1948; A new synthesis of glutamine and of γ-dipeptides of glutamic acid from phthalylated intermediates. *J Chem Soc, 1949:* 3315.
42. Sheehan, J. C., and Frank, V. S.: A new synthetic route to peptides. *J Am Chem Soc, 71:* 1856, 1949.
43. Schnabel, E., Herzog, H., Hoffmann, P., Klauke, E., and Ugi, I.: Synthese und Verwendung von *tert*-Butyloxycarbonylfluorid und anderen Fluorkohlensäureestern zur Darstellung säurelabiler Urethan-Derivate von Aminosäuren. *Liebigs Ann Chem, 716:* 175, 1968.

44. Carpino, L. A.: Oxidative reactions of hydrazines. IV. Elimination of nitrogen from 1,1′-disubstituted-2-arenesulfonhydrazides. *J Am Chem Soc, 79:* 4427, 1957.
45. Schnabel, E.: Verbesserte Synthese von *tert*-Butyloxycarbonyl-aminosäuren durch pH-Stat-Reaktion. *Liebigs Ann Chem, 702:* 188, 1967.
46. Schnabel, E., Stoltefuss, J., Offe, H. A., and Klauke, E.: Umsetzung von *tert*-Butyloxycarbonylfluorid mit trifunktionellen Aminosäuren. *Liebigs Ann Chem, 743:* 57, 1971.
47. Schnabel, E., Schmidt, G., and Klauke, E.: Neue Reagenzien zur Darstellung der [2-*p*-biphenyl-isopropyloxycarbonyl]-aminosäuren. *Liebigs Ann Chem, 743:* 69, 1971.
48. Miller, H. K., and Waelsch, H.: Benzyl esters of amino acids. *J Am Chem Soc, 74:* 1092, 1952.
49. Roeske, R. W.: Amino acid *tert*-butyl esters. *Chem Ind, 1959:* 1121.
50. Anderson, G. W., and Callahan, F. M.: *t*-Butyl esters of amino acids and peptides and their use in peptide synthesis. *J Am Chem Soc, 82:* 3359, 1960.
51. Hofmann, K., Magee, M. Z., and Lindenmann, A.: Studies on polypeptides. II. The preparation of α-amino acid carbobenzoxyhydrazides. *J Am Chem Soc, 72:* 2814, 1950.
52. Boissonnas, R. A., Guttmann, S., and Jaquenoud, P.-A.: Synthèse de la L-arginyl-L-prolyl-L-prolyl-glycyl-L-phenylalanyl-L-séryl-L-prolyl-L-phenylalanyl-L-arginine un nonapeptide présentant les propriétés de la bradykinin. *Helv Chim Acta, 43:* 1349, 1960.
53. Thomas, J. O.: The preparation of N-formyl derivatives of amino acids using N,N′-dicyclohexylcarbodiimide. *Tet Lett, 335:* 1967.
54. Nefkens, G. H. L.: Synthesis of phthaloylamino acids under mild conditions. *Nature, 185:* 309, 1960.
55. Panetta, C. A., and Casanova, T. G.: Trifluoroacetylation of amino acids and peptides under neutral conditions. *J Org Chem, 35:* 4275, 1970.
56. Sakakibara, S., and Shimonishi, Y.: A new method for releasing oxytocin from fully protected nonapeptides using anhydrous hydrogen fluoride. *Bull Chem Soc Japan, 38:* 1412, 1965.
57. Sakakibara, S.: The use of hydrogen fluoride in peptide chemistry. In Weinstein, B. (Ed.) : *Chemistry and Biochemistry of Amino Acids, Peptides and Proteins.* New York, Marcel Dekker, 1971, pp. 51–85.
58. Ben-Ishai, D., and Berger, A.: Cleavage of N-carbobenzoxy groups by dry hydrogen bromide and hydrogen chloride. *J Org Chem, 17:* 1564, 1952.
59. Guttmann, S., and Boissonnas, R. A.: Synthèse du N-acetyl-L-séryl-L-tyrosyl-L-séryl-L-méthionyl-γ-L-glutamate de benzyle et de peptides apparentés. *Helv Chim Acta, 41:* 1852, 1958.
60. Kappeler, H., and Schwyzer, R.: Die Synthese eines Tetracosapeptids mit der Aminosäuresequenz eines hochaktiven Abbauproduktes des

β-Corticotropins (ACTH) aus Schweinehypophysen. *Helv Chim Acta, 44:* 1136, 1961.

61. Hiskey, R. G., and Adams, Jr., J. B.: Sulfur-containing polypeptides. IV. Synthetic routes to cysteine peptides. *J Org Chem, 31:* 2178, 1966.
62. Sifferd, R. H., and du Vigneaud, V.: A new synthesis of carnosine, with some observations on the splitting of the benzyl group from carbobenzoxy derivatives and from benzylthio ethers. *J Biol Chem, 108:* 753, 1935.
63. Roberts, C. W.: The synthesis of L-cysteinyl-L-tyrosyl-L-isoleucine. *J Am Chem Soc, 76:* 6203, 1954.
64. du Vigneaud, V., Audrieth, L. F., and Loring, H. S.: The reduction of cystine in liquid ammonia by metallic sodium. *J Am Chem Soc, 52:* 4500, 1930.
65. Greenstein, J. P., and Winitz, M.: *Chemistry of the Amino Acids.* New York, Wiley & Sons, 1960, vol. II, p. 925.
66. Ontjes, D. A., and Anfinsen, C. B.: Solid-phase synthesis of a polypeptide sequence from staphylococcal nuclease. In Weinstein, B., and Lande, S. (Eds.) : *Peptides: Chemistry and Biochemistry.* New York, Marcel Dekker, 1970, pp. 79–98.
67. Veber, D. F., Milkowski, J. D., Denkewalter, R. G., and Hirschmann, R.: The synthesis of peptides in aqueous medium. IV. A novel protecting group for cysteine. *Tet Lett, 3057:* 1968.
68. Bergmann, M., Zervas, L., and Rinke, H.: Neues Verfahren zur Synthese von Peptiden des Arginins. *Hoppe-Seyler's Z Physiol Chem, 224:* 40, 1934.
69. Lenard, J.: Nitration and denitration in hydrogen fluoride. *J Org Chem, 32:* 250, 1967.
70. Curtius, T.: Synthetische Versuche mit Hippurazid. *Ber Dtsch Chem Ges, 35:* 3226, 1902.
71. Wieland, T., and Bernhard, H.: Uber Peptidsynthesen. 3. Die Verwendung von Anhydriden aus N-acylierten Aminosäuren und Derivaten anorganischer Säuren. *Liebigs Ann Chem, 572:* 190, 1951.
72. Boissonnas, R. A.: Une nouvelle méthode de synthèse peptidique. *Helv Chim Acta, 34:* 874, 1951.
73. Vaughan, Jr., J. R.: Acylalkylcarbonates as acylating agents for the synthesis of peptides. *J Am Chem Soc, 73:* 3547, 1951.
74. Sheehan, J. C., and Hess, G. P.: A new method of forming peptide bonds. *J Am Chem Soc, 77:* 1067, 1955.
75. Bodanszky, M.: Synthesis of peptides by aminolysis of nitrophenyl esters. *Nature, 175:* 685, 1955.
76. Kemp, D. S., Wang, S. W., Busby, III, G., and Hugel, G.: Microanalysis by successive isotopic dilution. A new assay for racemic content. *J Am Chem Soc, 92:* 1043, 1970.

77. Honzl, J., and Rudinger, J.: Nitrosylchloride and butyl nitrite as reagents in peptide synthesis by the azide method; suppression of amide formation. *Coll Czech Chem Commun, 26:* 2333, 1961.
78. Sieber, P., Riniker, B., Brugger, M., Kamber, B., and Rittel, W.: Menschliches Calcitonin. VI. Die Synthese von Calcitonin M. *Helv Chim Acta, 53:* 2135, 1970.
79. Schnabel, E.: Nebenreaktionen bei der Synthese von Peptiden nach dem Azidverfahren von Curtius. *Liebigs Ann Chem, 659:* 168, 1962.
80. Anderson, G. W., Zimmerman, J. E., and Callahan, F. M.: A reinvestigation of the mixed carbonic anhydride method of peptide synthesis. *J Am Chem Soc, 89:* 5012, 1967.
81. Kopple, K. D., and Renick, R. J.: Formation of an N-acylamide in peptide synthesis. *J Org Chem, 23:* 1565, 1958.
82. Zaoral, M., and Rudinger, J.: Product formed from tosylglycine under the conditions of a mixed carbonic anhydride synthesis. *Coll Czech Chem Commun, 26:* 2316, 1961.
83. Meienhofer, J., and du Vigneaud, V.: Preparation of lysine-vasopressin through a crystalline protected nonapeptide intermediate and purification of the hormone by chromatography. *J Am Chem Soc, 82:* 2279, 1960.
84. Kemp, D. S.: Mechanism of peptide coupling reactions. In Nesvadba, H. (Ed.): *Peptides 1971.* Amsterdam, North-Holland Co., 1973, pp. 1–19.
85. Wünsch, E., and Drees, F.: Zur Synthese des Glucagons. X. Darstellung der Sequenz 22–29. *Chem Ber, 99:* 110, 1966.
86. Weygand, F. Hoffmann, D., and Wünsch, E.: Peptidsynthesen mit Dicyclohexylcarbodiimid unter Zusatz von N-Hydroxysuccinimid. *Z Naturforsch, 21b:* 426, 1966.
87. König, W., and Geiger, R.: Eine neue Methode zur Synthese von Peptiden: Aktivierung der Carboxylgruppe mit Dicyclohexylcarbodiimid unter Zusatz von 1-Hydroxy-benzotriazolen. *Chem Ber, 103:* 788, 1970.
88. Sheehan, J. C., Goodman, M., and Hess, G. P.: Peptide derivatives containing hydroxyamino acids. *J Am Chem Soc, 78:* 1367, 1956.
89. Anderson, G. W., Zimmerman, J. E., and Callahan, F. M.: N-Hydroxysuccinimide esters in peptide synthesis. *J Am Chem Soc, 85:* 3039, 1963; The use of esters of N-hydroxysuccinimide in peptide synthesis. *J Am Chem Soc, 86:* 1839, 1964.
90. Pless, J., and Boissonnas, R. A.: Uber die Geschwindigkeit der Aminolyse von verschiedenen neuen aktivierten, N-geschützten α-Aminosäure-phenylestern, insbesondere 2,4,5-trichlorphenylestern. *Helv Chim Acta, 46:* 1609, 1963.
91. Kupryszewski, G., and Formela, M.: Amino acid chlorophenyl esters. III. *N*-Protected amino acid pentachlorophenyl esters. *Roczniki Chem, 35:* 1533, 1961. [*Chem Abstr, 57:* 7373, 1962.]

92. Jakubke, H.-D., and Voigt, A.: Untersuchungen über die peptidchemische Verwendbarkeit von Acylaminosäure-chinolyl-(8)-estern. *Chem Ber 99:* 2419, 1966.
93. Bodanszky, M., and du Vigneaud, V.: A method of synthesis of long peptide chains using a synthesis of oxytocin as an example. *J Am Chem Soc, 81:* 5688, 1959.
94. Bodanszky, M., Meienhofer, J., and du Vigneaud, V.: Synthesis of lysine-vasopressin by the nitrophenyl ester method. *J Am Chem Soc, 82:* 3195, 1960.
95. Hirschmann, R., and Denkewalter, R. G.: The synthesis of an enzyme. *Naturwissenschaften, 57:* 145, 1970.
96. Schwyzer, R., and Sieber, P.: Total synthesis of adrenocorticotrophic hormone. *Nature, 199:* 172, 1963.
97. Bajusz, S., Medzihradszky, K., Paulay, Z., and Láng, Zs.: Totalsynthese des menschlichen Corticotropins (α_h-ACTH). *Acta Chim Acad Sci Hung, 52:* 335, 1967.
98. Beacham, J., Dupuis, G., Finn, F. M., Storey, H. T., Yanaihara, C., Yanaihara, N., and Hofmann, K.: Fragment condensations with peptide derivatives related to the primary structure of ribonuclease T_1. *J Am Chem Soc, 93:* 5526, 1971.
99. Camble, R., Dupuis, G., Kawasaki, K., Romovacek, H., Yanaihara, N., and Hofmann, K.: Synthesis of a protected tritricontapeptide hydrazide corresponding to positions 48–80 of the primary structure of Ribonuclease T_1. *J Am Chem Soc, 94:* 2091, 1972.
100. Rittel, W., Brugger, M., Kamber, B., Riniker, B., and Sieber, P.: Thyrocalcitonin III. Die Synthese des α-Thyrocalcitions. *Helv Chim Acta, 51:* 924, 1968.
101. Guttmann, S., Pless, J., Sandrin, E., Jaquenoud, P.-A., Bossert, H., and Willems, H.: Synthese des Thyrocalcitonins. *Helv Chim Acta, 51:* 1155, 1968.
102. Sieber, P., Brugger, M., Kamber, B., Riniker, B., and Rittel, W.: Menschliches Calcitonin. IV. Die Synthese von Calcitonin M. *Helv Chim Acta, 51:* 2057, 1968.
103. Guttmann, S., Pless, J., Huguenin, R. L., Sandrin, E., Bossert, H., and Zehnder, K.: Synthese von Salm-Calcitonin, einem hochaktiven hypocalcämischen Hormon. *Helv Chim Acta, 52:* 1789, 1969.
104. Bodanszky, M., Ondetti, M. A., Levine, S. D., Narayanan, V. L., *von* Saltza, M., Sheehan, J. T., Williams, N. J., and Sabo, E. F.: Synthesis of a heptacosapeptide amide with the hormonal activity of secretin. *Chem Ind, 1966:* 1757.
105. Boissonnas, R. A., and Guttmann, S.: Chemistry of the neurohypophyseal hormones. In Berde, B. (Ed.): *Handbook of Experimental Pharmacology.* New York, Springer, 1968, vol. XXIII, pp. 40–66.
106. Bumpus, F. M., Smeby, R. R., and Khairallah, P. A.: Synthsis and biological properties of angiotensin II analogs. In Weinstein, B.,

and Lande, S. (Eds.): *Peptides: Chemistry and Biochemistry.* New York, Marcel Dekker, 1970, pp. 127–150.

107. Stewart, J. M., and Woolley, D. W.: The search for peptides with specific antibradykinin activity. In Erdös, E. G., Back, N., and Sicuteri, F. (Eds.): *Hypotensive Peptides.* New York, Springer, 1966, pp. 23–31.

108. Matsuura, S., Waki, M., Makisumi, S., and Izumiya, N.: Studies on peptide antibiotics. XVI. Analogs of gramicidin S containing β-alanine in place of L-proline. *Bull Chem Soc Jap, 43:* 1197, 1970.

109. Shemyakin, M. M., Vinogradova, E. I., Feigina, M. Y., Aldanova, N. A., Loginova, N. F., Ryabova, I. D., and Pavlenko, I. A.: The structure-antimicrobial relation for valinomycin depsipeptides. *Experientia, 21:* 548, 1965.

110. Finn, F. M., Visser, J. P., and Hofmann, K.: Structure-function studies in partially synthetic ribonuclease. In Bricas, E. (Ed.): *Peptides 1969.* Amsterdam, North-Holland, 1969, pp. 330–335.

111. Scoffone, E., Marchiori, F., Moroder, L., Rocchi, R., and Scatturin, A.: Structure and activity relationship in some partially synthetic modified ribonucleases. In Bricas, E. (Ed.): *Peptides 1969.* Amsterdam, North-Holland, 1969, pp. 325–329.

112. Sayers, M. A., Sayers, G., and Woodbury, L. A.: The assay of adrenocorticotrophic hormone by the adrenal ascorbic acid-depletion method. *Endocrinology, 42:* 379, 1948.

113. Boissonnas, R. A., Guttmann, S., and Pless, J.: Synthesis of D-Ser^1-Nle^4-$(Val\text{-}NH_2)^{25}$-β-corticotropin (1–25), a highly potent analogue of ACTH. *Experientia, 22:* 526, 1966.

114. Desaulles, P. A., Riniker, B., and Rittel, W.: High corticotrophic activity of a short-chain synthetic corticotrophin analogue. *Excerpta Medica Internat Congress Series, 161:* 489, 1968.

115. Geiger, R.: Synthese eines Heptadecapeptids mit hoher adrenocorticotroper Wirkung. *Liebigs Ann Chem, 750:* 165, 1971.

116. Kusch, P.: Synthese von Nylonoligomeren nach dem Merrifieldschen Verfahren der Peptidsynthese am Kunstharz. *Kolloid Z und Z Polymere, 208:* 138, 1966.

117. Anfinsen, C. B., Ontjes, D., Ohno, M., Corley, L., and Eastlake, A.: The synthesis of protected peptide fragments of a staphylococcal nuclease. *Proc Nat Acad Sci U S, 58:* 1806, 1967.

118. Takashima, H., du Vigneaud, V., and Merrifield, R. B.: The synthesis of deamino-oxytocin by the solid phase method. *J Am Chem Soc, 90:* 1323, 1968.

119. Merrifield, R. B.: Solid-phase peptide synthesis. II. The synthesis of bradykinin. *J Am Chem Soc, 86:* 304, 1964.

120. Lenard, J., and Robinson, A. B.: Use of hydrogen fluoride in Merrifield solid-phase peptide synthesis. *J Am Chem Soc, 89:* 181, 1967.

121. Marshall, G. R., and Merrifield, R. B.: Synthesis of angiotensins by the solid-phase method. *Biochemistry, 4:* 2394, 1965.
122. Merrifield, R. B.: New approaches to the chemical synthesis of peptides. In Pincus, G. (Ed.) : *Recent Progress in Hormone Research.* New York, Academic Press, 1967, vol. 23, pp. 451–482.
123. Khosla, M. C., Smeby, R. R., and Bumpus, F. M.: Apparatus for solid-phase peptide synthesis. *Science, 156:* 253, 1967.
124. Brunfeldt, K., Halstrom, J., and Roepstorff, P.: A punched tape controlled peptide synthesizer. *Acta Chem Scand, 23:* 2830, 1969.
125. Grahl-Nielsen, O., and Tritsch, G. L.: Synthesis of oligomeric L-lysine peptides by the solid-phase method. *Biochemistry, 8:* 187, 1969.
126. Birr, C., and Lochinger, W.: Ein neuer Reaktor für die Merrifield-Synthese erprobt an der Synthese eines Dekapeptids des Antamanids. *Synthesis, 1971:* 319.
127. Merrifield, R. B.: Automated synthesis of peptides. *Science, 150:* 178, 1965.
128. Merrifield, R. B.: The synthesis of biologically active peptides and proteins. *JAMA, 210:* 1247, 1969.
129. Merrifield, R. B., Stewart, J. M., and Jernberg, N.: Instrument for automated synthesis of peptides. *Anal Chem, 38:* 1905, 1966.
130. Sano, S., and Kurihara, M.: Synthesis of an analogue of horse heart cytochrome *c* by the solid phase method. *Hoppe-Seyler's Z Physiol Chem, 350:* 1183, 1969.
131. Izumiya, N., Waki, M., Kato., T., Ohno, M., Aoyagi, H., and Misuyasu, N.: Solid-phase synthesis of ribonuclease T_1. In Meienhofer, J. (Ed.) : *Chemistry and Biology of Peptides.* Ann Arbor, Ann Arbor Science Publ., 1972. pp. 269–279.
132. Hancock, W. S., Prescott, D. J., Nulty, W. L., Weintraub, J., Vagelos, P. R., and Marshall, G. R.: The synthesis of a protein with acyl carrier protein activity. *J Am Chem Soc, 93:* 1799, 1971.
133. Noda, K., Terada, S., Mitsuyasu, N., Waki, M., Kato, T., and Izumiya, N.: Synthesis of a peptide with basic pancreatic trypsin inhibitor activity. *Naturwissenschaften, 58:* 147, 1971.
134. Bayer, E., Jung, G., and Hagenmaier, H.: Untersuchungen zur Totalsynthese des Ferredoxins. I. Synthese der Aminosäuresequenz von *C. Pasteurianum* Ferredoxin. *Tetrahedron, 24:* 4853, 1968.
135. Ontjes, D. A., and Anfinsen, C. B.: Solid-phase synthesis of a 42-residue fragment of staphylococcal nuclease: properties of a semi-synthetic enzyme. *Proc Nat Acad Sci US, 64:* 428, 1969.
136. Potts, Jr., J. T., Tregear, G. W., Keutmann, H. T., Niall, H. D., Sauer, R., Deftos, L. J., Dawson, B. F., Hogan, M. L., and Auerbach, G. D.: Synthesis of a biologically active N-terminal tetratriacontapeptide of parathyroid hormone. *Proc Nat Acad Sci U S, 68:* 63, 1971.

137. Marglin, A., and Merrifield, R. B.: The synthesis of bovine insulin by the solid phase method. *J Am Chem Soc, 88:* 5051, 1966.

138. Blake, J., Wang, K.-T., and Li, C. H.: Adrenocorticotropin XLI. The solid-phase synthesis of α^{1-19}ACTH,alanyl-α^{1-19}-ACTH and prolyl-α^{1-19}-ACTH and their adrenocorticotropic activity. *Biochemistry, 11:* 438, 1972.

139. Blake, J., and Li, C. H.: The solid-phase synthesis of alpha-melanotropin. *Int J Protein Res, 3:* 185, 1971.

140. Wieland, T., Birr, C., and Flor, F.: Synthese von Antamanid mit der Merrifield-Technik. *Liebigs Ann Chem, 727:* 130, 1969.

141. Ohno, M., Kuromizu, K., Ogawa, H., and Izumiya, N.: An improved synthesis of gramicidin S *via* solid-phase synthesis and cyclization by the azide method. *J Am Chem Soc, 93:* 5251, 1971.

142. Li, C. H., Liu, W.-K., and Dixon, J. S.: Human pituitary growth hormone. XII. The amino acid sequence of the hormone. *J Am Chem Soc, 88:* 2050, 1966.

143. Li, C. H., Dixon, J. S., and Liu, W.-K.: Human pituitary growth hormone. XIX. The primary structure of the hormone. *Arch Biochem Biophys, 133:* 70, 1969.

144. Li, C. H., and Dixon, J. S.: Human pituitary growth hormone. XXXII. The primary structure of the hormone: revision. *Arch Biochem Biophys, 146:* 233, 1971.

145. Losse, G., and Neubert, K.: Peptidsynthese an hochpolymeren Verbindungen. *Z Chemie, 10:* 48, 1970.

146. Marshall, G. R., and Merrifield, R. B.: Peptides prepared by solid-phase peptide synthesis. In Sober, A. (Ed.): *CRC Handbook of Biochemistry, with Selected Data for Molecular Biology.* Cleveland, Chemical Rubber Co., 1970, pp. C145–C162.

147. Stewart, J. M., and Young, J. D.: *Solid Phase Peptide Synthesis.* San Francisco, Freeman, 1969.

147a. Bayer, E., Eckstein, H., Hägele, K., König, W. A., Brüning, W., Hagenmaier, H., and Parr, W.: Failure sequences in the solid phase synthesis of polypeptides. *J Am Chem Soc, 92:* 1735, 1970.

148. Yaron, A., and Schlossman, S. F.: Preparation and immunologic properties of stereospecific α-dinitrophenylnonalysines. *Biochemistry, 7:* 2673, 1968.

149. Wang, S.-S., and Merrifield, R. B.: *tert*-Alkyloxycarbonylhydrazide-resin and *tert*-alkyl alcohol-resin for preparation of protected peptide fragments. In Scoffone, E. (Ed.): *Peptides 1969.* Amsterdam, North-Holland, 1971, pp. 74–83.

150. Kenner, G. W., McDermott, J. R., and Sheppard, R. C.: The safety catch principle in solid phase peptide synthesis. *J Chem Soc Chem Commun 1971:* 636.

151. Wang, S.-S., and Merrifield, R. B.: Preparation of a *tert*-alkyloxycar-

bonylhydrazide resin and its application to solid phase peptide synthesis. *J Am Chem Soc, 91:* 6488, 1969.

152. Bayer, E., Jung, G., Halász, I., and Sebestian, I.: A new support for polypeptide synthesis in columns. *Tet Lett, 1970:* 4503.
153. Parr, W., and Grohmann, K.: A new solid support for peptide synthesis. *Tet Lett, 1971:* 2633.
154. Tregear, G. W., Niall, H. D., Potts, Jr., J. T., Leeman, S. E., and Chang, M. M.: Synthesis of substance P. *Nature New Biology, 232:* 87, 1971.
155. Meienhofer, J., and Trzeciak, A.: Solid-phase synthesis with attachment of peptide to resin through an amino acid side chain: [8-lysine]-vasopressin. *Pro Nat Acad Sci U S, 68:* 1006, 1971.
156. Yamashiro, D., Blake, J., and Li, C. H.: The use of N^{α}, N^{im}-bis *(tert-*butyloxycarbonyl) histidine and N^{α}-2- (*p*-biphenylyl) isopropyloxycarbonyl, histidine in the solid-phase synthesis of histidine-containing peptides. *J Am Chem Soc, 94:* 2855, 1972.
157. Wang, S.-S., and Merrifield, R. B.: Preparation of some new biphenylisopropyloxycarbonylamino acids and their application to the solid phase synthesis of a tryptophan-containing heptapeptide of bovine parathyroid hormone. *Int J Protein Res, 1:* 235, 1969.
158. Southard, G. L., Brooke, G. S., and Pettee, J. M.: Solid phase peptide synthesis. I. A mild method of solid phase peptide synthesis employing an enamine nitrogen protecting group and a benzhydryl resin as a solid support. *Tet Lett, 1969:* 3505.
159. Southard, G. L., Brooke, G. S., and Pettee, J. M.: Preparation of N-(2-benzoyl-l-lmethylvinyl) amino acid dicyclohexylammonium salts for peptide synthesis. *Tetrahedron, 27:* 1359, 1971.
160. Carpenter, F. H., and Gish, D. T.: The application of *p*-nitrobenzyl chloroformate to peptide synthesis. *J Am Chem Soc, 74:* 3818, 1952.
161. Dorman, L. C.: A non-destructive method for the determination of completeness of coupling reactions in solid phase peptide synthesis. *Tet Lett, 1969:* 2319.
162. Kaiser, E., Colescott, R. L., Bossinger, C. D., and Cook, P. I.: Color test for detection of free terminal amino groups in the solid-phase synthesis of peptides. *Anal Biochem, 34:* 595, 1970.
163. Wieland, T., Lüben, G., Ottenheym, H., Faesel, J., de Vries, J. X., Konz, W., Prox, A., and Schmid, J.: The discovery, isolation, elucidation of structure, and synthesis of antamanide. *Angew Chem, Int Ed, 7:* 204, 1968.
164. Wieland, T.: Poisonous principles of mushrooms of the genus amanita. *Science, 159:* 946, 1968.
165. Waki, M., and Izumiya, N.: Studies of peptide antibiotics. X. Syntheses of cyclosemigramicidin S and gramicidin S. *Bull Chem Soc Jap, 40:* 1687, 1967.

166. Kessler, W., and Iselin, B.: Selektive Spaltung substituierter Phenylsulfenyl-Schutzgruppen bei Peptidsynthesen. *Helv Chim Acta, 49:* 1330, 1966.

167. Ohno, M., and Anfinsen, C. B.: Removal of protected peptides by hydrazinolysis after solid-phase synthesis. *J Am Chem Soc, 89:* 5994, 1967.

168. Meienhofer, J., Jacobs, P. M., Godwin, H. A., and Rosenberg, I. H.: Synthesis of hepta-γ-L-glutamic acid by conventional and solid-phase techniques. *J Org Chem, 35:* 4137, 1970.

169. Mazur, R. H., Schlatter, J. M., and Goldkamp, A. H.: Structure-taste relationships on some dipeptides. *J Am Chem Soc, 91:* 2684, 1969; Structure-taste relationships of some small peptides. In Weinstein, B., and Lande, S. (Eds.): *Peptides: Chemistry and Biochemistry.* New York, Marcel Dekker, 1970, pp. 175–180.

170. Steiner, D. F., and Oyer, P. E.: The biosynthesis of insulin and a probable precursor of insulin by a human islet cell adenoma. *Proc Nat Acad Sci, U S, 57:* 473, 1967.

171. Steiner, D. F., Hallund, O., Rubenstein, A., Cho, S., and Bayliss, C.: Isolation and properties of proinsulin, intermediate forms, and other minor components from crystalline bovine insulin. *Diabetes, 17:* 725, 1968.

172. Chance, R. E., Ellis, R. M., and Bromer, W. W.: Porcine proinsulin: characterization and amino acid sequence. *Science, 161:* 165, 1968.

CHAPTER 2

PEPTIDES IN HEMOGLOBIN SYNTHESIS

HENRY BROWN AND JULIE BROWN

INTRAVENOUS AMINO ACIDS are often an excellent source for protein nutrition, but some very ill patients, for example with liver failure, do not use them well. (1) Intravenous whole protein as in blood or plasma must usually be used for these patients. One probable explanation for at least part of the difficulty is that some peptide bonds cannot be synthesized easily by some very ill patients. Such patients might use amino acids very well if they contained these peptide linkages, as in whole blood protein.

Because of this problem, for a number of years we have been studying the possibility of incorporating amino acids in peptide linkage into one blood protein, hemoglobin. This paper is a progress report in this study.

In the late 1950's we reported use of amino acids in peptide linkage from a peptic hydrolysate of globin for new hemoglobin synthesis. (2–5) More amino acids in peptide linkage were incorporated into new hemoglobin than were corresponding amounts of free amino acids. (3,6) In addition, specific activity of the same amino acid varied in different peptides from a peptic hemoglobin hydrolysate. Incorporation of preformed peptide intermediates into new hemoglobin peptide chains was proposed as the most likely explanation. Both alpha and beta (α and β) peptide chains

This work was aided in part by U.S. Public Health Service Grant No. AM-08681 might incorporate these peptides. Possibly only one chain used these peptides or one chain was synthesized from them in preference to the other chain. Because of these possibilities, in the first part of this report experiments concerning separation and rate of synthesis of rat hemoglobin α and β chains will be discussed. The second part of the report will document synthesis of one possible decapetide intermediate for rat hemoglobin. Evidence will be reviewed for and against incorporation of preformed peptides into newly forming hemoglobin.

SEPARATION AND SPECIFIC ACTIVITIES OF AND CHAINS OF RAT HEMOGLOBIN

Rat hemoglobin containing ^{14}C labeled glycine and serine was obtained from their red cells after injecting rats with 40 μC ^{14}C glycine intraperitoneally daily for four days (160 μC total) as previously described. (2–5) α and β chains were isolated from 0.3 gm of this labeled globin with a specific activity of 0.002982 μC per mg using Wilson and Smith's method. (7) Steps of separation are detailed since separation of chains of this specific hemoglobin have not been reported to our knowledge. One hundred grams of Amberlite GC 50 type 2 resin was washed first with 4 N sodium hydroxide, then with 4 N hydrochloric acid followed by water and finally 11.7 percent formic acid with which the resin was permitted to equilibrate. A 40 cm by 2.2 cm column of resin was poured. The 0.3 gm of globin was dissolved in 20 ml of 11.7 percent formic acid and shaken with 30 ml of the remaining formic acid equilibrated resin. A few mg of formic acid dissolved globin precipitated when resin was added. Dissolved globin and resin were poured into the resin column increasing its height to 46.5 cm. The column was washed with 165 ml of 2 M urea. A gradient elution was done beginning with 2 M urea in a 2 liter beaker into which 8 M urea from a one liter beaker was siphoned and stirred. The urea in gradually increasing concentration was pumped through the column at a flow rate of 2 ml per minute. Twelve and five-tenths ml fractions were collected by a Gilson Fraction Collector.

These 3 liters were pumped through the column followed by an additional liter of 8 M urea. α and β chains were identified by absorbance readings of aliquots at 280 mμ on a Beckman model DB spectrophotometer. Five hundred and fifty ml of eluate containing the α chain and 625 ml containing the β chain were freed from urea by dialyzing in cellophane sacks against running tap water for 3 days. Each was then dialyzed against 8 liters of distilled water for 2 additional days. Precipitation of α and β chains occurred with removal of urea. Water was decanted after transferring frac-

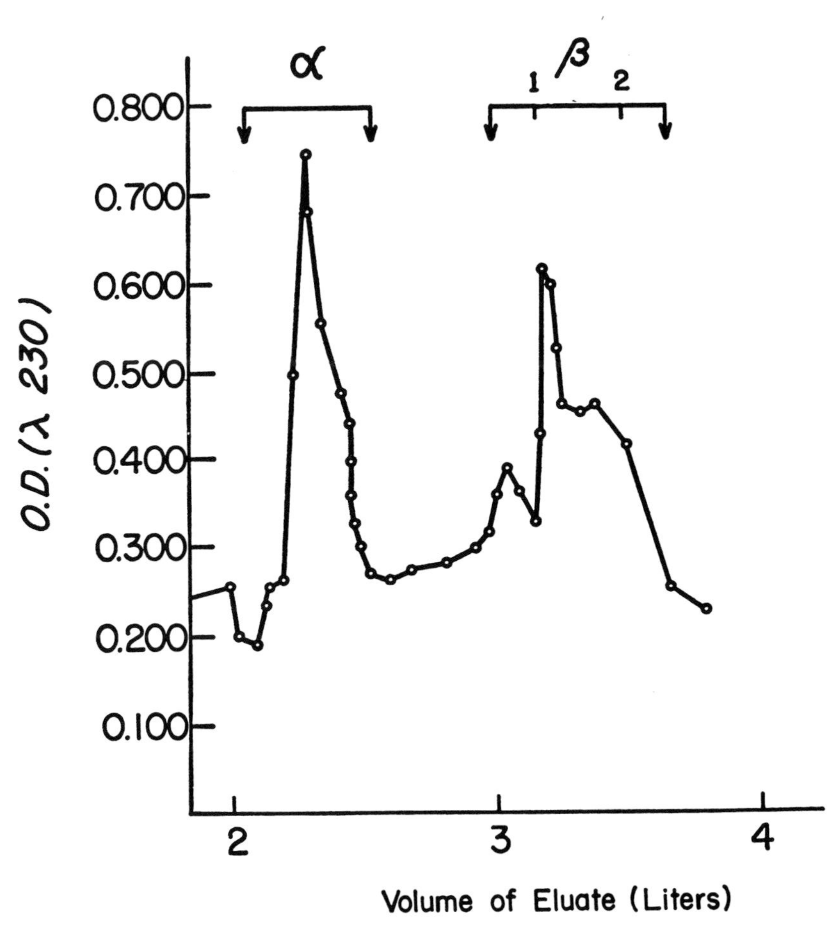

Figure 2-1. Elution chromatogram of α and β chains at hemoglobin.

tions to separate beakers. Forty ml of acetone were added to each precipitate. They were refrigerated overnight. After centrifuging each suspension and decanting acetone, α and β chains were each dried in a vacuum desiccator over sulfuric acid and sodium hydroxide.

Sixty-three mg of α and 10 mg of β chain were obtained or 42 percent and 7 percent theoretic yields respectively. The elution chromatogram (Figure 2-1) suggested that rat globin contained at least one lesser component in the α fraction and at least two lesser components in the β fraction. This is consistent with the findings of others (28).

Following the β chain, two additional peaks were eluted, but not shown in Figure 2-1. Curves drawn from an absorption spectrum scan from 400 to 200 mμ of the material taken from the first of these two additional peaks was similar to that of the β chain itself. This suggested the possibility of β chain-like material in the peak eluted following the β chain.

AMINO ACID COMPOSITION AND SPECIFIC ACTIVITIES OF ^{14}C GLYCINE AND SERINE IN α AND β CHAINS

Four mg of α and 4 mg of β chains were separately dissolved in 6 drops of 5.7N hydrochloric acid at 37 degrees centigrade. Each sample was then sealed in a capillary tube and hydrolyzed for 24 hours at 105 degrees centigrade. After removal of hydrochloric acid by evaporation an amino acid analysis of an aliquot of each was done with a Beckman amino acid analyzer (Table 2-I).

Radioactivity was determined in a second aliquot of each hydrolysate. Previous studies showed that over 90 percent of radioactivity in the globin was in glycine and serine. (3) Specific activity of glycine and serine counted together was approximately 16 percent greater in β than in α chains after four days (Table 2-II). Significance of this relatively small difference is not clear.

SYNTHESIS OF PEPTIDE 13

Glycine, serine and leucine of rat globin were labeled with ^{14}C as described for gylcine above. Peptide 13 (Figure

TABLE 2-I.

Amino Acid Composition of α and β Chains of Rat Hemoglobin

Amino Acids	*α Chain* *uM/mg*	*β Chain* *uM/mg*
Lysine	0.496	0.475
Histidine	0.397	0.361
Arginine	0.117	0.126
Cysteic Acid	0.202	0.040
Aspartic Acid	0.661	0.475
Threonine	0.179	0.137
Serine	0.204	0.123
Glutamic Acid	0.336	0.268
Proline	0.244	0.198
Glycine	0.498	0.362
Alanine	0.639	0.486
Methionine	0.044	0.031
Isoleucine	0.127	0.115
Leucine	0.645	0.490
Tyrosine	0.075	0.070
Phenylalanine	0.250	0.225
Valine	0.482	0.410

2-2) was selected for synthesis because it was the most radioactive of the 26 studied from a peptic globin hydrolysate (Table 2-III). In addition, specific activities of the three labeled amino acids were greater than found when each was measured from the entire globin molecule (Table 2-IV). Amino acid sequence of peptide 13 was:

C-leu.ser.val.ser.ala.gly.leu.asp.glu.ser-N (6)

The Merrifield solid phase method used for synthesis of this decapeptide is well known in principle. (8,9,11) Problems in this specific synthesis, nevertheless, indicated need for the following detailed discussion.

Resin Preparation

Five grams of copolymers of styrene 2 percent divinyl benzene resin coupled through its carboxyl group to t-butyloxycarbonyl-leucine (Boc-leucine) was sequentially at-

TABLE 2-II.

Specific Activity of Glycine and Serine in α and β Chains of Rat Hemoglobin

	α	β
cpm/uM Glycine and Serine	3,060	3,600

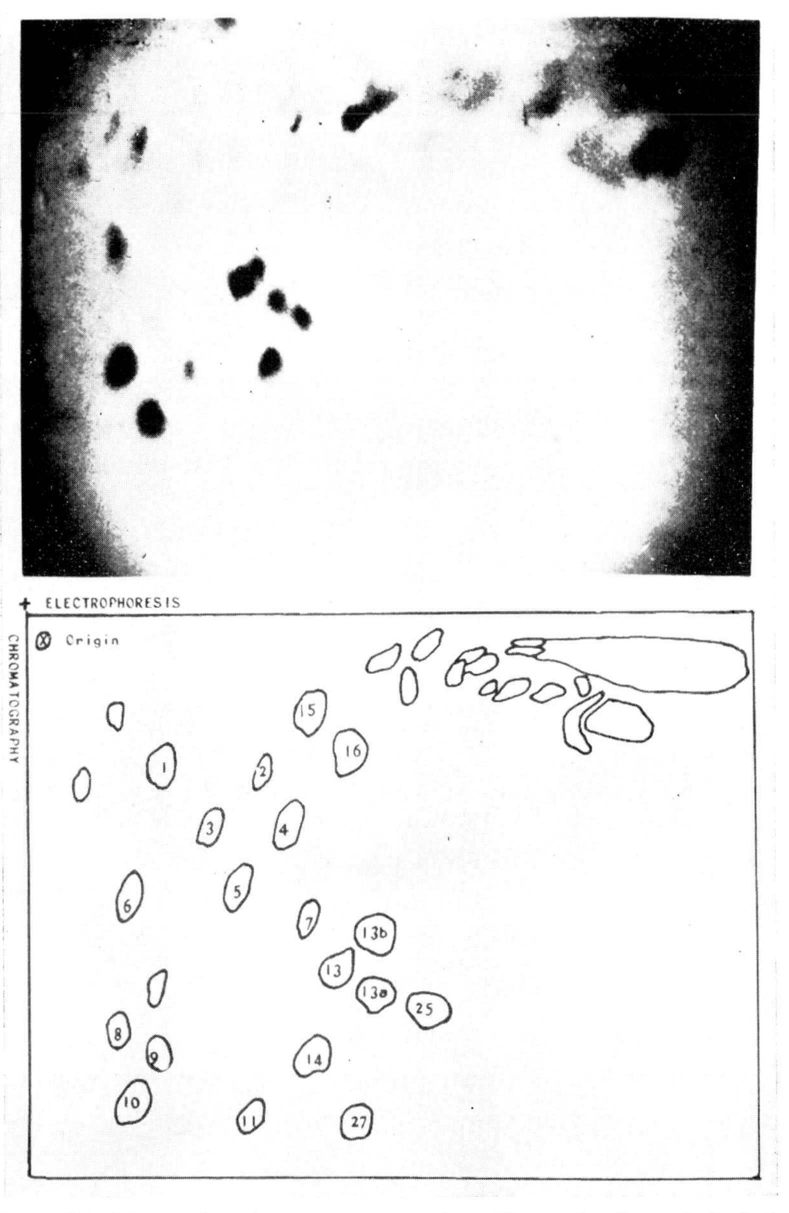

Figure 2-2. Electrophoresis chromatogram (peptide map) of peptic hydrolepate of rat globin (from Brown[6]).

TABLE 2-III.

Relative radioactivity of 26 peptides from a peptic hydrolysate of 12.5 mg rat globin labeled with ^{14}C serine, ^{14}C glycine and ^{14}C leucine

Peptide Number	*CPM*	*Peptide Number*	*CPM*
1	4.7	14	97.4
2	20.3	15	36.3
3	8.7	16	46.8
4	38.7	17	93.3
5	14.1	18	204.5
6	2.8	19	254.5
7	30.1	20	61.8
8	0	21	118
9	35.8	22	23.3
10	26	23	79.3
11	201.2	24	82.6
12	120.9	25	63.7
13	448.5	26	25.8
13a	264		

tached to the other Boc amino acids. The resin was placed in a medium-sized Merrifield vessel (an 80 × 30 mm cylinder) containing a sintered glass filter at the bottom.

Reagent Preparation

Volumes of reagents used were 10 ml per gram of resin and washings were for three minutes each. All reagents were purified by redistillation and used within a few days. Methylene chloride and chloroform were both filtered through a 3 × 4 cm aluminum oxide (Al_2O_3) column after distillation to remove traces of hydrochloric acid. Even though hydrochloric acid is present in low concentrations, appreciable amounts are accumulated in the large volumes of solvents used in washing. These amounts of hydrochloric acid could

TABLE 2-IV

Specific activities of ^{14}C glycine, serine and leucine in peptide 13 and in the entire rat hemoglobin molecule (from Brown, H.[6]).
Unequal Distribution of Radioactivity in Labeled Rat Globin

	Counts per Minute / μ Mole	
	Globin	*Peptide 13*
Serine	4,000	9,800
Glycine	4,500	8,100
Leucine	9,250	26,500

(From Brown, H.[6])

be very damaging to the relatively small amount of peptide on resin. Storage was in dark bottles. Molecular sieve #4 was also added to methylene chloride for further purification.

To prepare 1N hydrochloric acid in acetic acid, gaseous hydrochloric acid was first dried by passing it through sulfuric acid. The gas was then dissolved in acetic acid until correct weight was reached.

Boc-leucine Resin Preparation

Boc-leucine resin must be further purified prior to peptide synthesis. It was washed three times with methylene chloride, three times with ethanol, three times with glacial acetic acid and finally, three more times with methylene chloride. Resin was allowed to rock overnight in the Merrifield machine in the methylene chloride wash. This relatively long period allowed swelling or opening of resin so that reactive groups were available. Between washings, solvents were filtered off with a water aspirator being careful to avoid passing air through the resin.

Cycle for Peptide Synthesis

Each cycle of the synthesis of the peptide consisted of addition of one amino acid by the following steps:

Deprotection

The N-terminal boc protective group was removed (deprotected) by rocking in 30 percent trifluoroacetic acid (TFA), in methylene chloride for five minutes, filtering and then rocking resin in TFA again for 25 minutes. Resin was washed with methylene chloride-glacial acetic acid in a ratio of 1:1. This was followed by 3 acetic acid washes. Deprotecting continued with two additions of glacial acetic acid containing 1N hydrochloric acid for five minutes and twenty-five minutes each. Resin was then washed 3 times with acetic acid, followed by acetic acid chloroform, 1:1 and finally, with three chloroform washes.

Neutralization

Ten percent triethylamine in chloroform was added for 10 minutes. This step was repeated once followed by chloroform and one methylene chloride chloroform, 1:1 wash. Filtrates from all washes were evaporated in a round bottom flask for a Volhard silver titration procedure for completeness of reaction discussed below. The cycle continued with three additional washings with methylene chloride (25).

Peptide Bonding

The preparation was then ready to be joined to the next amino acid by peptide linkage. Three equivalents of the next Boc amino acid were added in 45 ml methylene chloride to the resin for 20 minutes. This was not filtered. Three equivalents of dicyclohexylcarbodiimide (DCCI) were added in 5 ml methylene chloride and the resin rocked overnight. Following filtration the following washes were done: 3 with methylene chloride, 1 with methylene chloride with chloroform 1:1, 3 with chloroform and one with methylene chloride: chloroform, 1:1 and finally, 3 with methylene chloride. Before filtering the last methylene chloride wash, a small amount (0.5 ml) was pipetted from the reaction vessel for the Kaiser test. (10) This test is a sensitive ninhydrin test for detecting semi-quantitatively free amino groups. To the 0.5 ml of the final methylene chloride wash solution 3 drops of 3 reagents were added: (1) 5 percent ninhydrin (triketohydrindene hydrate) in 95 percent ethanol; (2) 80 grams of phenol dissolved in 20 ml ethanol; (3) 2 ml of 0.001M potassium chloride diluted to 100 ml with pyridine. The test was negative if resin beads remained colorless and solution yellow. Positivity of test was determined by various shades of blue of resin and solution. If positive, 3 additional equivalents of Boc amino acid and DCCI were to be added to the reaction vessel to couple with remaining free amino groups. During synthesis of this peptide, this latter step was not needed. When the Kaiser test was negative, the cycle began anew with cleavage of the Boc group from the last amino acid attached to the chain.

Volhard Test for Completeness of Reaction

The test measures equivalents of free amino groups indirectly by quantitatively determining chloride released from each amino group hydrochloride during neutralization with triethylamine. Residues of evaporated filtrates from neutralization were dissolved in 20 ml of 2N nitric acid. Fifteen ml of 0.1N silver nitrate were added to quantitatively precipitate chlorides. Volume was diluted to exactly 100 ml with 0.4N nitric acid (25). The solution was filtered through filter paper, discarding the first 20 ml.

Five-tenths ml of ferric ammonium sulfate solution, $Fe(NH_4)(SO_4)_2 \cdot 12H_2O$, was added as indicator to exactly 20 ml of the filtrate. The solution was titrated with 0.1N potassium thiocyanate solution until a red-brown color appeared. The test served both as an index of losses and completeness of reaction. If the number of free amino groups was very deficient, it was necessary to repeat deprotective steps.

Amino Acid Analysis for Purity of Peptide

This analysis is made at any point in synthesis by boiling 10 mg of resin peptide under reflux in 5.7N hydrochloric acid for 72 hours. Since hydrolysis on resin is very slow, for complete cleavage of amino acids, resin is filtered and hydrochloric acid removed by evaporation. The peptide is redissolved in 5.7N hydrochloric acid and heated at 100 degrees in an evacuated sealed tube for an additional 20 hours. Hydrochloric acid is removed by evaporation a second time and an amino acid analysis done with an aliquot. These tests, Kaiser, Volhard and amino acid analyses, for completeness of reaction and purity are useful for small peptides. As size increases beyond the range of decapeptides, they are insufficiently sensitive. Merits and pitfalls of these tests are considered in detail in Chapter One by Dr. Meienhofer.

Cleavage From Resin of the Peptide Without Hydrolysis

Several methods involving use of anhydrous hydrogen bromide or hydrogen fluoride were used. Anhydrous hydrogen fluoride (HF) was preferred and one successful method with this reagent evolved by Robinson and well described by Stewart and Young was as follows. (12,12a) The procedure was done in a good hood. One gram of resin containing 0.142 mM of peptide was placed in a translucent 100 ml plastic flask of kel-f, trifluorethylene, which is inert to HF. 0.3 ml anisole was added to the resin. The flask was placed in a solid carbon dioxide (dry ice) alcohol bath in a Duwar flask. Nitrogen was passed into the resin through a system of Teflon (tetrafluorethylene) fittings connected by kel-f tubing from kel-f rods. After approximately 30 minutes, nitrogen was discontinued. HF gas was then passed through the system for an additional one-half hour with constant stirring. The liquid volume increased to about 10 ml during this time. HF gas was turned off but stirring was continued for another one-half hour. At the end of this time, the vessel was slowly warmed to 37 degrees centigrade in a water bath. Nitrogen was bubbled through the resin a second time until all the liquid, i.e. HF and anisole, were gone. Resin was deep blue in the presence of HF and red-orange with HF and anisole. Red-orange color disappeared as HF and anisole, drawn off through potassium hydroxide by a water aspirator were completely removed. Peptide was washed from resin twice, each time with 10 ml of 90 percent acetic acid. Resin swells and solubility of peptide increases in this concentration of acetic acid. Resin was filtered and filtrate was lyophilized twice. Fluffy precipitate was dissolved in 0.1N acetic acid and placed on an 80 $\times$ 2 cm sephedex G 15 column to remove all components of molecular weight less than 700. The eluate containing the peptide was lyophilized, dissolved in 30 ml water and lyophilized again.

The white fluffy peptide was then dissolved in 15 ml 0.2N acetic acid and placed a second time on the same sephedex column. The 0.2N acetic acid eluate was collected in 80 drop aliquots. Peptides were identified by reading optical density at 230 mμ on a Beckman Du model spectrophotometer. A sharp peak occurred at fractions 18–28 (Figure 3). Fractions 18–21, #1, the ascending limb, were combined and lyophilized separately from fractions 22–27, #2, the descending limb. Fraction 1 weighed 7 mg and fraction 2 weighed 130 mg. A peptide map was made from each of these fractions on Whatman filter paper #3mm. Two methods of filter paper electrophoresis were used. One method used 0.2N acetic acid, pH 2.7, 1,700 volts D.C. for 2 hours for one dimension and chromatography in butanol acetic acid water 5:4:1 for the other dimension. Alternatively, electrophoresis was done at pH 2.1 in 0.1N formate-acetate buffer under toluene at 2,500 volts for 1 hour, cooling with running tap water. Chromatography for the second dimension was the same as

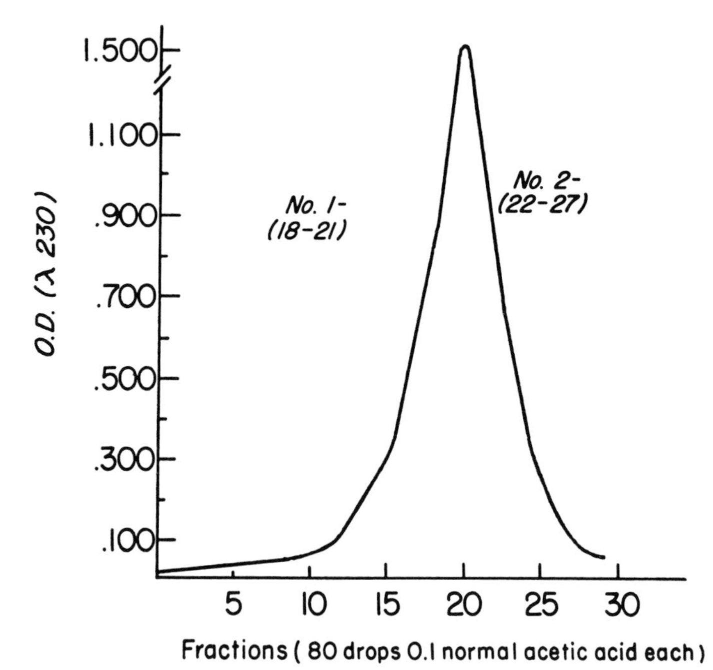

Figure 2-3. Synthetic rat hemoglobin peptide 13 elution chromatogram from Sephedex G 15.

above. Results with either type electrophoresis were similar. Peptide spots were identified by dipping papers in 0.01N ninhydrin dissolved in acetone. After standing 12 hours, peptide spots were cut out for elution. Papers were dipped in 0.1N ninhydrin for any spots not previously detected. The ascending limb, #1, had 5 and #2 had 9 peptide spots. A spot on each had the same R_f as the natural peptide and on elution, hydrolysis and quantitative amino acid analysis had the correct composition for peptide 13. The ascending limb, #1, was fractionated a second time on a sephedex G 10 column, eluting with 0.1N acetic acid. After lyophilizing fractions at the peak, 4.4 mg were obtained. An electrophoresis chromatogram was done. Similar treatment was given the descending limb. The electrophoresis chromatograms from each limb yielded one large spot. Peak fractions had an additional faint spot and the descending limb had three additional faint spots. The large main spot in each case had the correct amino acid analysis and serine as the N-terminal amino acid as determined with 1 fluoro-2,4 dinitrobenzene, FDNB, Sanger's reagent, (23) and Dansyl techniques. (24) Remaining ninhydrin positive spots on an electrophoresis chromatogram of each purified limb gave only trace amounts of amino acids. These quantities were minute enough to be contaminants of sephedex or filter paper indicating that the synthetic peptide was chemically pure.

Natural and synthetic peptides again both have the same R_f values on electrophoresis chromatograms made on Whatman filter paper #3 mm with 0.2N acetic acid in one direction and butanol acetic acid water in other as above.

Discussion

Human hemoglobin is believed to be synthesized in developing red blood cells "de novo" from amino acids at the rate of about 8 grams per day. Human globin contains two identical halves, each composed of two peptide chains termed alpha and beta (α and β). Formula weight of the

574 amino acids comprising all four chains is 61,992 and with heme attached to form completed hemoglobin, 64,458. (13)

A proposed position for natural and synthetic peptides 13 compared to human globin is shown in Figure 2-4. Peptide 13 probably comes from the alpha chain corresponding in position to amino acids 129–138 in human globin. Amino acid composition of these two sequences is compared in Figure 2-5. An acidic octapeptide with C- terminal leucine has, in fact, been isolated from this region from peptic hydrolysates of human chain. (21) If peptide 13 configuration corresponds to human chain positions 129–138, it would have an alpha-helical form in the segment conventionally labeled H. (22)

Further study of amino acid sequences of rat hemoglobin is particularly needed. Variations are reported in sequences of hemoglobin peptides within this species with more than one type of hemoglobin being reported even in the same rat. (28) This may explain at least in part why some suggest that sequence 129–138 is the same in both human and rodent hemoglobin. (29)

Early studies of rat hemoglobin containing ^{14}C labeled valine by Muir, Neuberger and Perrone showed N-terminal valine to have the same specific activity as valine in the remaining molecule. (14) From these data, they concluded that N-terminal valine elsewhere in the molecule were incorporated at approximately the same time. They theorized that free amino acids were rapidly condensed sequentially or else small preformed peptide intermediates were involved.

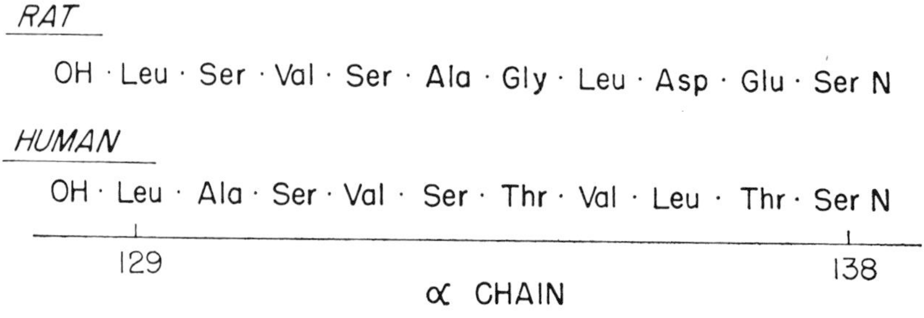

Figure 2-4. Proposed position of rat hemoglobin peptide 13 in α chain as compared to human hemoglobin chain.

AMINO ACID COMPOSITION SEQUENCE:
129-138 α CHAIN

	RAT	HUMAN
Leu	2	2
Ser	3	3
Val	1	2
Ala	1	1
Gly	1	0
Asp	1	0
Glu	1	0
Thr	0	2

Figure 2-5. Amino acid composition of peptide 13 from rat hemoglobin as compared to its proposed counterpart in human hemoglobin. (21)

Itano estimated 90 seconds as time required for synthesis of one hemoglobin molecule, each red cell completing 800 molecules per second. (15) Dintzis pointed out, however, that rate of chain growth is not uniform. (16) Pulse labeling techniques were used as follows. Bone marrow erythrocyte precursor cells were labeled in vitro for relatively long periods with ^{14}C lysine and for relatively short periods with 3H leucine. Results suggested that synthesis slows in the region of the ninetieth amino acid of α and β chains. This was the area of attachment of heme. Further, when ^{14}C labeled amino acids were administered as part of the incubation solution specific activity of amino acids in α chains early in the synthesis, the first 10 minutes, was greater than those of β chains. (17) After 10 minutes, specific activities in α and β chains were the same. Interpretation was that a small pool of α chains were synthesized prior to β chains. Another theoretic explanation is that different parts of peptides were labeled at different times. Relatively fewer labeled amino acids may have been available for peptides synthesized later around position 90. Alternatively, since heme and globin are not necessarily synthesized at the same time (27), heme may not be available at precisely the right time for linkage to amino acid 90 area. Chain prior to and beyond 90 might be made at approximately the same rate. The heme attaching peptides may be made later when heme is available. This difference in time of synthesis for parts of α and β chains is also analogous to entire alpha and beta chains being synthesized at different times.

Other evidence may be cited to support the concept of peptide intermediates. Whipple's idea of large segments of proteins moving from one kind of protein molecule to another, particularly in blood, (18) is not entirely tenable, particularly because amino acid sequences of these proteins are so different. On the other hand, a middle road is possible. Small peptides may be freely interchangeable between protein molecules. For example, albumin synthesis is often impaired in liver disease. Free amino acids administered intravenously in similar proportions to those found in albumin often are not well used. (1) Yet, whole albumin administered intravenously to these same patients is well utilized to serve many purposes, including synthesis of other proteins. Such data suggests that certain albumin peptide bonds are difficult for diseased liver to synthesize, but once formed, these linkages can be used for many purposes. Absorption of intact di- and tri-peptides (19) through gut wall has been reported. (20)

Trace amounts of peptides are also reported which exert enormous physiologic control. An example is control of thyroid secretion by Pyro-Glu. His. Pro. (26) Peptides may be present only fleetingly as other metabolites such as adenosine triphosphate or carbamyl phosphate, all difficult to detect for this reason but nonetheless present and important.

Finally, aliquots of the same globin peptides used well by

the whole animal in synthesizing new hemoglobin were not efficiently used for hemoglobin synthesis by reticulocytes when incubated *in vitro.* (5) Since most experiments of others involving hemoglobin synthesis discussed here were *in vitro,* (16,17) part of the discrepancy between those results and our *in vivo* experiments may be explained on that basis.

Two requirements must be met for proof of incorporation of the synthetic labeled peptide 13 into new globin. (1) The peptide must be shown to be labeled in the same manner in new hemoglobin with ratios or radioactivity of the amino acids to each other remaining the same. (2) Most of the radioactivity found in newly synthesized globin must be in the same sequence as that administered. If, for example, the peptide were broken down into individual amino acids before re-use, one would expect to find radioactive amino acids more uniformly throughout the molecule.

Summary

Separation of α and β chains of rat hemoglobin and their amino acid composition is discussed. Evidence is consistent with an unequal rate of synthesis of α and β chains over a 4-day period. Amino acids within the hemoglobin molecule have different specific activities at different places in the peptide chain. One decapeptide having different specific activity from the remaining molecule is probably from the α chain, amino acid positions 129–138. Hemoglobin regeneration experiments suggest that this entire decapeptide or part of it may be incorporated into new rat hemoglobin without prior breakdown to free amino acids. The synthesis of this peptide is detailed. The classical hypothesis for sequential synthesis of hemoglobin is reviewed. As an alternative or additional mechanism, hemoglobin synthesis from intermediates is proposed.

REFERENCES

1. Dudrick, S. J., McFayden, B. V., Van Buren, C. T., Ruberg, R. L., and Maynard, A. T.: Parenteral hyperalimentation: Metabolic problems and solutions. *Ann Surg, 176:* 259–264, Sept., 1972.

2. Brown, H., and Brown, J.: Utilization of peptides in hemoglobin formation. *Metabolism, 8:* 286–288, 1959.
3. Brown, H., and Brown, J.: Hemoglobin peptides used in hemoglobin synthesis. *Metabolism, 9:* 587–593, 1960.
4. Brown, H., and Brown, J.: Hemoglobin synthesis in hemorrhagic anemia. *JAMA, 179:* 143–145, 1962.
5. Brown, H., and Brown, J.: Utilization of Globin peptides in hemoglobin synthesis *in vitro. Metabolism, 10:* 91–93, 1961.
6. Brown, H.: *Hepatic Failure.* Springfield, Ill., Thomas, 1970, pp. 33–56.
7. Wilson, S., and Smith, D. B.: Separation of the valyl-leucyl and valyl-glutamyl polypeptide chains of horse globin by fractional precipitation and column chromatography. *Can J Biochem Physiol, 37:* 405–416, 1959.
8. Merrifield, R. B.: Solid phase peptide synthesis. *Adv Enzymol, 32:* 221, 1969.
9. Merrifield, R. B.: Automated synthesis of peptides. *Science, 150:* 178, 1965.
10. Kaiser, E., Colescott, R. L., Bossinger, C. D., and Cook, P. I.: Color test for detection of free terminal amino groups in the solid-phase synthesis of peptides. *Anal Biochem, 34*: 295, 1970.
11. Merrifield, R. B.: Solid phase peptide synthesis. I. The synthesis of a tetrapeptide. *J Am Chem Soc, 85:* 2149–2154, 1963.
12. Robinson, Thesis, University of California at San Diego, 1967, quoted by Stewart, J. M., and Young, J. D.: Solid phase peptide synthesis. W. H. Freeman, San Francisco, 1969.

12a. Lenard, J., and Robinson, A. B.: Use of HF in Merrifield solid phase peptide synthesis. *J Am Chem Soc, 89:* 181, 1967.

13. Harris, J. W., and Kellermeyer, R. W.: *The Red Cell.* Cambridge, Mass., Harvard, 1970, pp. 149–161.
14. Muir, H. M., Neuberger, A., and Perrone, J. C.: Further isotopic studies on hemoglobin formation in the rat and rabbit. *Biochem J, 52:* 87–95, 1952.
15. Itano, H. A.: Genetic regulation of peptide synthesis in hemoglobins. *J Cell Physiol, 67:* Sup., 65–75, 1966.
16. Dintzis, H. M.: Assembly of the peptide chains in hemoglobin. *J Nat Acad Sci, US, 17:* 247–261, 1961.
17. Winslow, R. M., and Ingram, V. M.: Peptide chain synthesis of human hemoglobins A and A_2. *J Biol Chem, 241:* 1144–1149, March, 1966.
18. Whipple, G. H.: The Dynamic equilibrium of body proteins. Springfield, Illinois, Thomas, 1956.
19. Eagle, H.: Utilization of dipeptides by mammalian cells in tissue culture. *Proc Soc Exp Biol Med, 89:* 96–99, 1955.
20. Mathews, D. M., and Laster, L.: Absorption of protein digestion products: A review. *Gut, 6:* 411–426, 1965.

21. Konigsberg, W., and Hill, R. J.: The structure of human hemoglobin. *J Biol Chem, 237:* 3157–3162, Oct., 1962.
22. *Br Med Bull, 25:* 14, 1969 quoted by Harris, J. W., and Kellermeyer, R. W.: *The Red Cell.* Cambridge, Mass., Harvard, 1970, p. 158.
23. Sanger, F.: The free amino groups of insulin. *Biochem J, 39:* 507, 1945.
24. Woods, K. R., and Wang, K. T.: Separation of dansyl-amino acids by polyamide layer chromatography. *Biochem et Biophys Acta, 133:* 369–370, 1967.
25. Hawk, P. B., Oser, B. L., and Summerson, W. H.: Chloride analysis by modified Volhard method. *P•actical Physiological Chemistry,* Blakiston, Philadelphia, 15th Ed., 1954, p. 955. Quoted in Stewart, J. M., and Young, J. D. *Solid Phase Peptide Synthesis,* W. H. Freeman, San Francisco, 1969.
26. Hershman, J. M., and Pittman, J. A.: Control of thyrotropin secretion in man. *New Eng J Med., 285:* 997–1005, (Oct. 28) 1971.
27. Drabkin, D. L., and Wise, C. D.: Independent synthesis of hemin and globin in hemoglobin. *Science, 132:* 1491, 1960.
28. Travnicek, T., Vodrazka, Z., Borova, J., Cejka, J., Salak, J., and Sulc, K.: The structural basis of polymorphism of rat hemoglobin. *Physiol Bohemoslov, 16:* 543–547, 1967.
29. Brdicka, R., Massa, A., Carta, S., Tentori, L., and Vivaldi, G.: Partial amino acid sequence of some tryptic petides of the alpha, chain of Rattus norvegicus hemoglobin. *Life Sciences II,* Part II: 895–899, 1972.

CHAPTER 3

AMINO ACIDS AND WHOLE PROTEIN UTILIZATION IN GERMFREE ANIMALS

STANLEY M. LEVENSON AND ELI SEIFTER

PROFOUND NUTRITIONAL EFFECTS of overt sepsis are obvious. What has not been so obvious are profound nutritional effects on mammals of their indigenous microbial flora. Consideration that this might be so has been voiced ever since microbes were discovered. Pasteur, (1) Nuttal and Thierfelder, (2) Schottelius, (3) Cohendy, (4) and Kuster (5) were among the early investigators concerned with such problems. One will recall that Metchnikoff (6) gathered evidence which he interpreted as indicating that certain colonic bacteria were harmful to human health. He believed other species of colonic bacteria as certain lactobacilli were helpful. The great technologic improvements in maintaining animals "germfree" or associated with specific, known microbial species were introduced by Glimstedt, (7) Reyniers and Trexler, (8) Miyakawa, (9) and Gustafsson. (10) Their specific work was also related to steel isolators for housing animals. Trexler (11) also introduced plastic isolators. All of

A number of our colleagues participated in many of these studies. They are: Drs. Erving F. Geever, Bud Tennant, Ernst Jaffe, Miss Dorinne Kan, Mr. Charles Gruber, Mr. Leo V. Crowley, Drs. Arnold Nagler, Giuseppe Rettura, Komei Nakao, Meir Lev, Ole Malm, Richard E. Horowitz, Floyd Doft, Llewellyn Ashburn, and Hyman Rosen.

We also appreciate the technical assistance of Mr. Alvin Watford and Mr. Raymond Alexander.

these studies laid the basis for an insurgence of experimental studies related to effects of indigenous microbial agents on mammalian metabolism and nutrition. (12,13,14,15,16) Important also were earlier studies of refection, including those of Eijkmann, (17) Cooper, (18) Fridericia, (19) and Coates. (20) Effects of coprophagy and its prevention were studied by Osborne and Mendel, (21) Steenbock, (22) and Barnes. (23) Briggs, (24) Moore, (25) Mameesh and Johnson, (26) Doft *et al.*, (27) Nielsen and Elvehjem, (28) Mickelsen, (29) Jukes, (30) Michel, (31) and many others studied effects of feeding various chemotherapeutic agents and antibiotics. In recent years, the studies of Dubos, Schaedler, and their associates (32,33,34) have contributed substantially to our understanding of this field. They investigated mice in protected, though not "germfree" environments and with limited indigenous microflora.

It has been long known that under some circumstances, the animal host is dependent on its intestinal microorganisms for synthesis of certain micronutrients. In other instances, intestinal microorganisms may increase dietary requirements of the animal host for certain micronutrients. In Table 3-I are listed some dietary nutritional deficiencies which have been studied in germfree (GF), conventionalized and conventional animals.* It is evident that in most instances the presence of indigenous microflora either aggravates or alleviates the dietary deficiency.

The classic example of the ameliorating effect of the microflora is in dietary *vitamin K* deficiency. Many strains of ordinary laboratory rats can go almost indefinitely without dietary vitamin K, while germfree rats promptly develop fatal hypoprothrombinemia and bleeding. (35,36) The reason is that the rat practices coprophagy. Certain gut bacteria (35) produce enough vitamin K, which is excreted in feces and then ingested to satisfy the rat's needs.

* Conventionalized animals are littermates of germfree animals that are purposefully contaminated with the cecal contents of conventional rats at weaning (21–23 days of age) and then maintained in isolators, using the same husbandry techniques as for germfree animals. Conventional animals are open animal room animals.

TABLE 3-I

Influence of Indigenous Microorganisms on Mammalian Nutritional Deficiencies

Aggravated	*Lessened*	*No Apparent Effect*
Vitamin C	Vitamin K	
Riboflavin	Thiamin	
Vitamin A	Folic Acid	
Vitamin E	Pantothenic Acid	
Selenium		
Cystine		
Choline (Renal)	Choline (Cirrhosis)	Choline (Early Liver Fat)
Amino Acids (e.g., Methionine)	Starvation	
Niacin		

In contrast, when guinea pigs are placed on a *vitamin C* deficient diet, the conventionalized guinea pig develops scurvy much sooner than the germfree guinea pig (37) (Fig. 3-1). This is evident from changes in food intake, body weight, gross and histologic lesions, changes in certain organ and blood ascorbic acid levels (Table 3-II), and death. (38)

A likely reason for differences in responses of conventionalized and germfree guinea pigs is "destruction" or "utilization" of ascorbic acid by intestinal flora of conventionalized guinea pigs. (39) It is likely that there is a constant movement of ascorbic acid from tissues and blood into the gut and back again. In the case of germfree guinea pigs, ascorbic acid can return unchanged, whereas in conventionalized guinea pigs, part of the ascorbic acid would be "destroyed" in the gut and thus, unavailable for recycling to other tissues and organs. Another possibility is that germfree guinea pigs may have a lower metabolic rate than conventionalized guinea pigs, as is the case for rats. (40)

Dietary *choline* deficiency appears unique in that it may be worsened, lessened, or unaffected by the presence of the indigenous microflora, depending on acuteness and severity of dietary deficiency and age of experimental animals, specifically rats. Thus, *the acute nephropathy* which occurs in very young rats on a choline-deficient diet is accentuated in the conventional rat as compared with the germfree. (41,

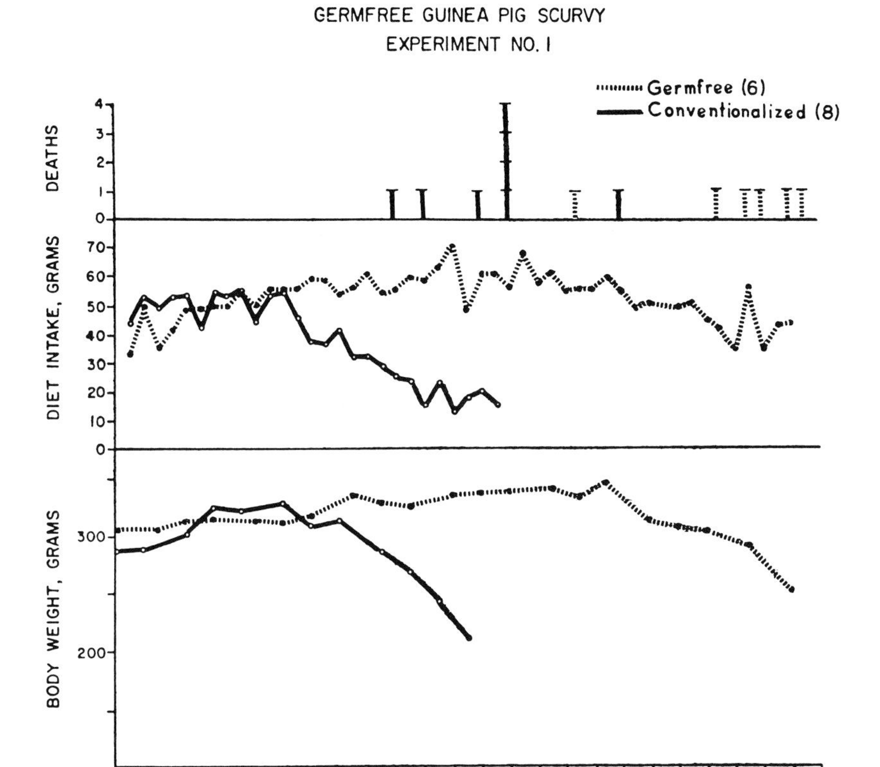

Figure 3-1. Body weights, food "intakes" (includes food eaten and scattered), and times of deaths of germfree and "conventionalized" guinea pigs receiving an ascorbic-acid-free diet. (Reproduced through the courtesy of *Arch Int Med, 110:* 693, 1962.)

42,43) The converse is so for *chronic liver cirrhosis* (44) (Tables 3-III and 3-IV).

Clearly, these differences in response to choline deficiency require more complicated explanations than those offered in the cases of vitamin K and vitamin C deficiencies. These latter explanations are simply either production of or utilization of a critical nutrient by indigenous microflora. The following are factors influencing action of choline. (1) It is true that choline is degraded to trimethylamine, a non-

TABLE 3-II

Effect of 7 Days of Ascorbic Acid Deprivation on Tissue and Blood Ascorbic Acid* Concentrations, Germfree and Conventionalized Guinea Pigs

	Liver (μg/g)		*Adrenal* (μg/g)		*Blood* (mg%)	
Days	*0*	*7*	*0*	*7*	*0*	*7*
CONV	230	63	1414	36	0.6	0.2
GF	217	222	1190	1299	0.6	0.4

* Includes ascorbic acid and dehydroascorbic acid.

lipotropic agent by the gut bacteria (45,46) (Fig. 3-2), but this is not a sufficient explanation for the varying effects of choline deficiency in germfree and conventional rats. In fact, a complex of interactions among a number of additional factors also are involved. These include, for example, (2) synthesis of one or more nutrients, such as vitamin B_{12}, by certain intestinal bacteria. These nutrients can ameliorate choline deficiency. (47,48) (3) In addition, other nutrients may be "destroyed," such as methionine. (49) It will

TABLE 3-III

The Effects of Acute Choline Deficiency* on Germfree, Conventionalized and Open Animal Room (OAR) Rats

				Non-Fatal Nephropathy			
No.	*Status*	*No. Rats*	*Fatal*	+++	++	+	*O*
I	OAR Water	11	10	1	0	0	0
J	Conv. Water	11	6	3	2	0	0
K	GF Water	11	0	0	0	0	11

* Acute choline deficiency induced by feeding of modified Salmon and Newberne diet (110), diet B.

Fischer rats, males, 21 days old at start of experimental diets. Survivors killed 12 days later. (Reproduced through courtesy of *J Nutr, 95(2):* 247–270, 1968.)

TABLE 3-IV

Dietary Liver Cirrhosis, Rats

	Germfree	*Conventionals*	
Exp. No.	*Isolator Mean grade**	*Isolator Mean grade**	*Open Animal Room Mean grade**
1	6.3 (7)	0.25 (8)	1.5 (12)
2	9.3 (9)	2.3 (15)	
4	5.4 (12)	1.3 (13)	

In each experiment, GF vs CONV, $p < 0.01$; no significant difference between conventionals housed in isolators and those housed in open animal room.

( ) No. of rats.

* Mean grade: Severity of liver cirrhosis graded 0–12; arbitrary grading system of Pathologist, L. Ashburn.

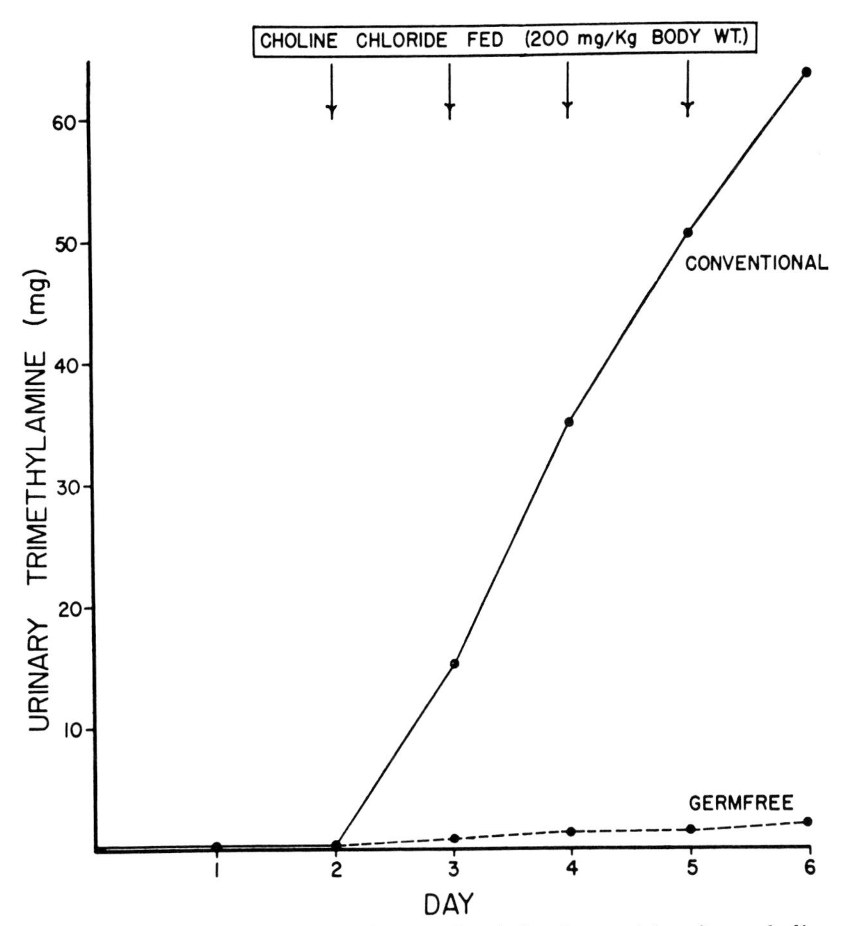

Figure 3-2. Cumulative total urinary trimethylamines arising from choline fed on four successive days to conventional and germfree rats. (Reproduced through the courtesy of *Arch Biochem Biophys, 94:* 424, 1961.)

be recalled that methionine can alleviate dietary choline deficiency. Other factors include (4) higher maintenance of hepatic choline synthesis by germfree rats than by conventional rats on choline deficient diets. (50) (5) Vasculature of germfree rats is less sensitive to pressor agents. (51) (6) Renal acetylcholine levels decrease more rapidly in conventional rats on choline deficient diets (52,53) (Tables 3-V and 3-Va) as compared with germfree rats. (7) Pos-

TABLE 3-V

Effects of Choline Deficiency on Open Animal Room Rats

Exp. No.	*Animal Status (Supplement)*	*Tissue*	*Acetyl Choline-Like Material*		*Acetyl Cholinesterase*	
			$\mu g/g$		μ moles acetyl choline hydrolyzed/g tissue/hr	
2 (7)	Water	Kidney		0.08 ± 0.03		15.9 ± 1.2
	Choline	Kidney	$P < 0.001$	0.27 ± 0.02	$P < 0.002$	19.5 ± 0.5
3 (6)	Water	Kidney		0.07 ± 0.04		15.0 ± 1.3
	Choline	Kidney	$P < 0.001$	0.28 ± 0.03	NS	19.4 ± 2.6
4 (10)	Water	Kidney		0.12 ± 0.03		18.6 ± 1.2
	Choline	Kidney	$P < 0.001$	0.23 ± 0.03	$P < 0.05$	22.4 ± 1.4

Fischer rats, 21 day old males at start of diets; killed 6 days later.
(Reproduced through courtesy of *J Nutrition, 94: (1):* 13–19, 1968.)

TABLE 3-Va

Tissue levels of acetylcholine and acetylcholinesterase during choline deficiency

Tissue	*Open-animal-room rats*[1]						*Germfree rats*[2]					
	Acetylcholine			*Cholinesterase*			*Acetylcholine*			*Cholinesterase*		
	Choline[3] *(8)*[4]	*Water (8)*		*Choline (8)*	*Water (8)*		*Choline (7)*	*Water (7)*		*Choline (7)*	*Water (7)*	
	$\mu g/g$	$\mu g/g$	*P*	$\mu g/hr/g$[5]	$\mu g/hr/g$[5]	*P*	$\mu g/g$	$\mu g/g$	*P*	$\mu g/hr/g$	$\mu g/hr/g$	*P*
Small intestine	0.79 ± 0.04[6]	0.54 ± 0.03	< 0.02	232 ± 23	219 ∓ 12	ns[7]	0.82 ∓ 0.03	0.63 ∓ 0.04	> 0.02	195 ∓ 13	190 ∓ 9	ns
Kidney	0.26 ± 0.02	0.16 ± 0.03	< 0.02	....	....	....	0.24 ± 0.02	0.27 ± 0.02	ns	....	....	....

[1] Open-animal-room, male rats of the Fischer strain, 20–22 days old at the start of the diet; rats were killed 5 days later.

[2] Germfree male rats of the Fischer strain, 20–22 days old at the start of the diet; rats were killed 5 days later.

[3] Choline chloride, 1.5 mg/ml in drinking water.

[4] Numbers in parentheses = numbers of rats. The experiment was conducted with 24 open-animal-room rats and 20 germfree rats; 4 of the 12 choline-deficient and 4 of the 12 choline-supplemented open-animal-room rats, and 3 of the 10 choline-deficient and 3 of the 10 choline-supplemented germfree rats were killed for histological examination of the kidneys. The remaining rats were killed for the measurements of acetylcholine and acetylcholine-esterase in the tissues.

[5] Acetylcholine, μg, hydrolyzed per hour per gram tissue.

[6] Mean ± SE.

[7] ns = not significant.

(Reproduced through courtesy of *J Nutrition, 95 (4):* 603–606, 1968.)

sibly conventional rats have a greater level of pressor agents because of production of certain amines by action of gut bacteria. Factors 5, 6, and 7 are important, we think, particularly in regard to development of the renal lesion in acute choline deficiency of young rats. This lesion is less severe in germfree rats. Finally, (8) the lower metabolic rate of germfree rats compared with conventional rats would result in certain aspects of choline deficiency in the germfree being less severe.

Our thesis regarding the pathogenesis of acute choline deficiency nephropathy is that a disturbance in renal blood flow is fundamental. We believe that in acute choline deficiency, decreased levels of acetylcholine lead to vasospasm of renal vasculature because of altered reactivity to pressor agents. Renal ischemia, necrosis, and hemorrhage follow. In support of this hypothesis are our studies showing (1) an abrupt drop in kidney acetylcholine levels in rats eating a choline-deficient diet. (2) Reactivity of the circulation of the meso-appendix of such rats is altered in a manner consistent with an acute acetylcholine deficiency. (54) (3) Certain non-lipotopic cholinergic drugs such as β-methyl choline and pyridyl choline (55) minimize or prevent the nephropathy of acute choline deficiency.

We have offered the following hypothesis to explain differences between choline deficient germfree and conventional rats in developing liver cirrhosis. Germfree are more susceptible to cirrhosis than conventional rats, in contrast to the lessened susceptibility of conventional rats treated with certain antibiotics. (56,57,58) We assume that in conventional rats some intestinal bacteria synthesize an "anti-cirrhotic nutrient (s) ." Other bacteria do not make this nutrient. These organisms normally compete with one another. We postulate that when certain antibiotics are fed, those bacteria not producing the "anti-cirrhotic nutrient" are reduced. Those which make the postulated nutrient now increase and the nutrient they produce is available in significantly greater amounts to the animal. As reported, therefore, onset of cirrhosis is delayed. In contrast, germfree animals have *no* viable bacteria and, therefore, none of the "protective nutrient" is synthesized by intestinal bacteria. It follows that germfree rats may develop cirrhosis more rapidly than conventional rats as, indeed, they did.

We have long considered the possibility that the postulated nutrient may be vitamin B_{12}. Certain intestinal bacteria synthesize vitamin B_{12}, and such B_{12} is available to the rat, generally through coprophagy. As mentioned, vitamin B_{12} will prevent development of liver cirrhosis by open-animal room rats on low-choline (47,48) diets. This is also so for germfree rats. (43)

The lower metabolic rate of the germfree rats would also lessen the adverse effects of chronic choline deficiency, as it would for acute choline deficiency. (59)

The beneficial effect of the germfree state on the acute hemorrhagic nephropathy of choline deficiency contrasts with its effect on development of liver cirrhosis. Similarly, neomycin given to open animal room weanling rats on choline deficient diets had no effect on development of the acute nephropathy or early accumulation of liver fat. Later, however, accumulation of liver fat and onset of cirrhosis were delayed while severity of cirrhosis was lessened. Reasons for this paradox regarding acute and chronic choline deficiency appears as follows. Lesser severity of acute choline deficiency in germfree rats reflects both "utilization" of the specific nutrients, choline and methionine as well as increased metabolic rate caused by certain intestinal bacteria of conventional rats. On the other hand, greater severity of chronic choline deficiency for germfree rats reflects synthesis of small amounts of a "protective" nutrient, such as vitamin B_{12}, by intestinal bacteria. The amount of choline required daily to prevent the acute nephropathy is much less than that required to prevent liver fat accumulation and liver cirrhosis. We postulate that neomycin does not change the intestinal bacterial flora rapidly enough to influence acute changes which occur in a matter of days. In time, however, bacterial flora does change under the influence of neomycin. More of the protective nutrient (s) is synthesized and made available to the rat. Development of liver cirrhosis is thereby delayed or prevented.

As mentioned, it is our thought that certain bacteria are effective in preventing low choline dietary liver cirrhosis while others are ineffective, or may be antagonistic. This may be the basis for known variations in response of different strains of rats to dietary choline deficiency liver cirrhosis. Patek, Defritsch, and Hirsch (60) have speculated ". . . that inherited traits in some unknown fashion exert an important effect on the susceptibility of the rat to dietary cirrhosis." We suggest that variations in intestinal microflora among different strains of rats may also be important in this regard.

Oxygen Consumption, Carbon Dioxide, Body Temperature

We have already mentioned that germfree rats have 15 to 20 percent lower oxygen consumption than their conventional or conventionalized counterparts. This is also true for carbon dioxide production (40) (Table 3-VI). The fasting RQ is the same for germfree and conventional rats. When values of oxygen consumption and carbon dioxide production were expressed per kilogram body weight minus gut content weight, these values were still significantly higher for the conventionalized. These calculations were made because of significantly larger weight of gut contents of germfree rats. Although amounts of body fat in germfree and conventionalized rats differed, amounts of carcass, gut, liver, and total body protein did not. Differences in oxygen consumption and carbon dioxide production are not due to differences in lean body mass. Colonic temperatures of germfree rats were significantly lower, about 1° centigrade, than those of conventionalized.

In subsequent experiments, we found that oxygen consumption and carbon dioxide production in rats harboring only *Cl. welchii* or *Bacteroides sp.* did not differ from their germfree littermates. These parameters were significantly higher in rats monocontaminated with *E. coli* whether values were expressed per kilogram body weight or per kilogram

TABLE 3-VI

Influence of Microorganisms on Oxygen Consumption, Carbon Dioxide Production, Colonic Temperature, Body Weight and Gastrointestional Tract Weight of Rats[1]

Exp. no.	No. of rats	Microbial status	Age	Body wt	GI tract empty	GI tract contents	Colonic temperature	Before correction for GI contents: O_2 consumption	Before correction for GI contents: CO_2 production	After correction for GI contents: O_2 consumption	After correction for GI contents: CO_2 production	RQ
			months	g	g	g	degrees	liters/kg body wt per day		liters/kg body wt per day		
	6	CONV[2]	4.5	190 ± 14[3]	7.5 ± 0.5	7.7 ± 0.8	37.6 ± 0.1	25.5 ± 0.3	22.8 ± 0.4	26.6 ± 0.4	23.7 ± 0.7	0.89 ± 0.0
1	6	GF[2]	4.5	225 ± 26	10.9 ± 0.7	21.9 ± 1.2	36.5 ± 0.2	20.9 ± 0.4	18.0 ± 0.6	23.1 ± 0.4	19.9 ± 0.4	0.86 ± 0.0
		P		ns	0.01	0.001	0.02	0.001	0.005	0.005	0.02	ns
	4	CONV	3	149 ± 17	5.3 ± 0.2	3.9 ± 0.7	37.0 ± 0.1	25.3 ± 0.4	21.6 ± 0.2	25.9 ± 0.4	22.2 ± 0.2	0.8 ± 0.0
2A	4	GF	3	168 ± 13	5.8 ± 0.2	22.9 ± 2.2	36.4 ± 0.1	20.4 ± 0.5	17.3 ± 0.2	23.6 ± 0.4	20.1 ± 0.2	0.8 ± 0.0
		P		ns	ns	0.001	0.02	0.005	0.005	0.05	0.01	ns
	4	*Cl. welchii*	2.5	127 ± 9.9	5.5 ± 0.3	12.0 ± 1.1	37.3 ± 0.1	25.8 ± 0.3	22.9 ± 0.5	28.2 ± 0.5	25.3 ± 0.7	0.89 ± 0.0
2B	4	GF	2.5	139 ± 5.0	5.9 ± 0.1	17.8 ± 1.4	36.7 ± 0.3	25.2 ± 0.1	22.1 ± 0.6	28.9 ± 0.4	25.4 ± 0.2	0.88 ± 0.0
		P		ns	ns	0.02	ns	ns	ns	ns	ns	ns
	6	*E. coli*	3	175 ± 9.0	7.7 ± 0.4	12.3 ± 1.1	37.1 ± 0.1	25.4 ± 0.3	22.2 ± 0.7	27.4 ± 0.5	23.9 ± 0.7	0.87 ± 0.0
2C	6	GF	3	191 ± 14.2	7.0 ± 0.4	14.2 ± 1.1	36.8 ± 0.1	22.8 ± 0.2	19.9 ± 0.3	24.6 ± 0.3	21.4 ± 0.2	0.87 ± 0.0
		P		ns	ns	ns	ns	0.01	0.05	0.01	0.02	ns

[1] Experiments 1, 2A, 2B: each group consisted of equal numbers of male and female Fischer rats. In experiment 2C, all rats were females.
[2] CONV = conventionalized; GF = germfree.
[2] Mean ± SE.

(Reproduced through courtesy of *J Nutrition, 97 (4)*: 542–552, 1969.)

body weight minus gut contents. Gastrointestinal tract size and contents are only slightly smaller in *E. Coli, Cl. welchii* and *Bacteroides sp.* rats than in germfree rats. Respiratory quotients were similar in all groups. Colonic temperatures of germfree, *Cl welchii, Bacteroides sp.,* and *E. coli* rats were similar, while that of the conventionalized was higher

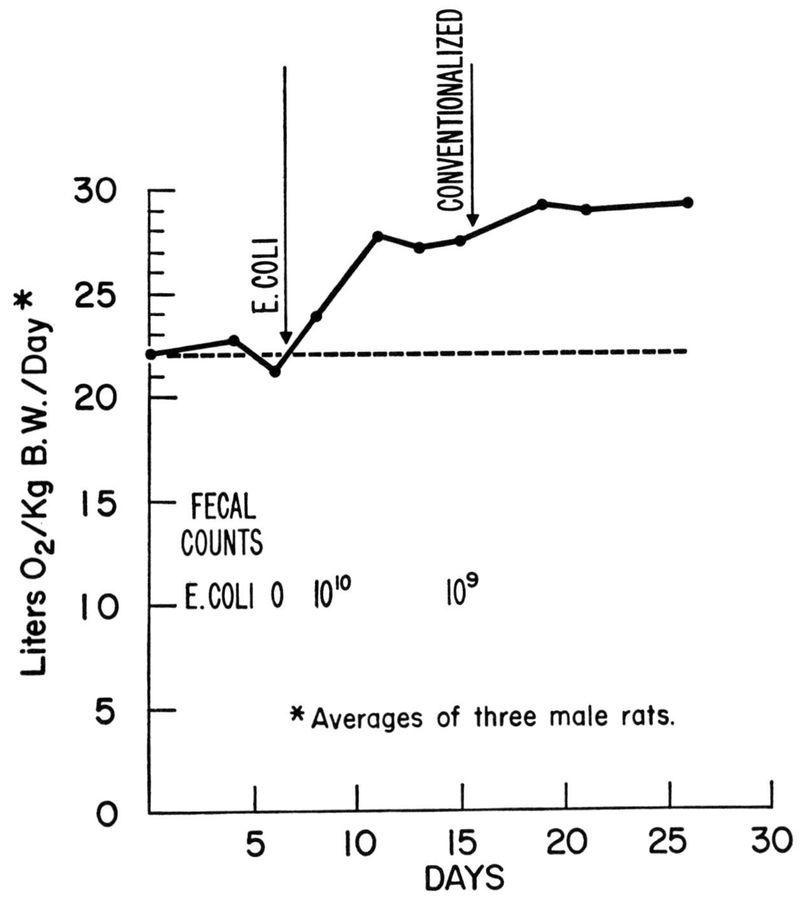

Figure 3-3. Effects of purposeful association of rats with *E. coli* followed by conventionalization of the rats with cecal contents of conventional rats.

than that of their germfree controls. All findings were independent of sex.

Increased O_2 consumption and CO_2 production after purposeful contamination with *E. coli* occurred very promptly indeed, at least as early as 12 hours, but lagged somewhat behind the number of viable *E. coli* found in feces. The peak effect on metabolic rate was seen in 3 to 4 days. There was a modest further rise in O_2 production when the *E. coli* rats were conventionalized (Fig. 3-3).

Germfree rats fed enormous numbers of heat-killed *E. coli* in their drinking water showed no increase in oxygen consumption or carbon dioxide production. When these same rats were fed viable *E. coli*, however, the number of viable *E. coli* in the gut, 10^9/gm feces 24 hours after inoculation, and oxygen consumption and carbon dioxide production promptly increased. The respiratory quotient did not change. A culture of *Proteus sp.* was fed to the *E. coli* rats in an attempt to increase the number of facultative anaerobes in the gut to see how this would modify oxygen consumption and carbon dioxide production. Rats were tested 5 and 7 days later. We did not achieve our objective since total number of viable bacteria in feces did not change demonstrably. *E. coli* and *Proteus* were each present in 10^8 viable bacteria/gm feces. Oxygen consumption and carbon dioxide production did not change (Fig. 3-4).

Neomycin was then given to these rats in amounts of 0.7 mg/ml drinking water daily. Five and seven days later, the number of viable organisms per gram of feces had dropped, *E. coli*, 10^4–10^6, Proteus, 10^6. There was a concomitant drop in oxygen consumption and carbon dioxide production.

Neither dead *E. coli*, viable *Bacteroides, E. coli,* or *Proteus*, nor neomycin changed the colonic temperatures of the rats.

We found, however, in a number of studies carried out during several years that in all but one experiment, colonic temperature of germfree rats was significantly lower than that of conventionalized. In the one exception, measurements were made after rats had been conventionalized for

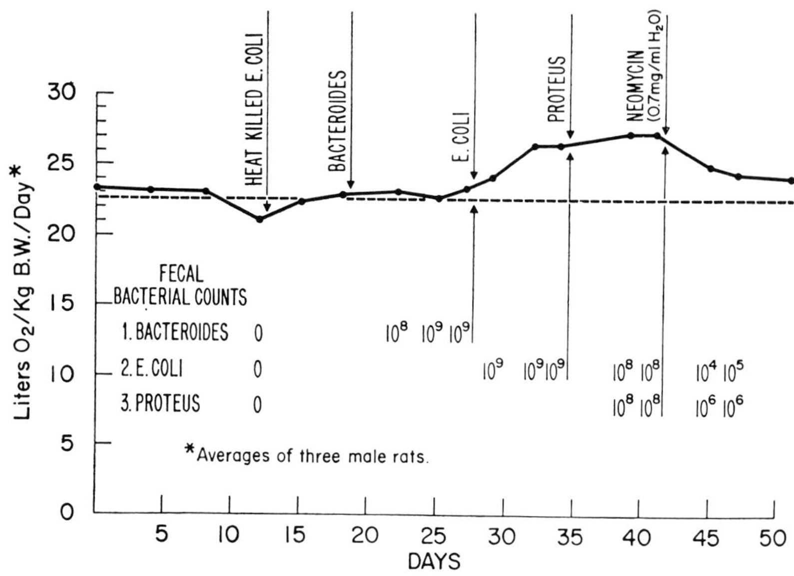

Figure 3-4. Oxygen consumption of germfree rats and following the sequential feeding of heat-killed *E. coli,* viable Bacteroides sp., *E. coli,* Proteus sp. and neomycin. Bacterial counts expressed as number of viable bacteria per gram feces. (Reproduced through courtesy of *J Nutr, 97:* 542, 1969.)

two weeks. In the others, rats had been conventionalized for a minimum of six weeks.

Influence of Intestinal Bacteria on Serum PBI and Serum Thyroxine Iodine Concentrations

Serum protein bound iodine (PBI) and serum thyroxine iodine concentrations were similar in germfree, *E. coli,* monocontaminants, *Bacteroides* monocontaminants, and conventionalized rats. These observations were made two weeks after purposeful contaminations at a time when differences in oxygen consumption were evident (Table 3-VII).

Influence of Specific Bacteria on the Turnover of Intestinal Mucosal Epithelial Cells

We speculated that differences in turnover rates of intes-

TABLE 3-VII

Influence of Intestinal Bacteria on Serum PBI, and Thyroxine Iodine of Rats

Microbial Status	*PBI* (*μg/100 ml serum*)	*Thyroxine Iodine*[1] (*μg/100 ml serum*)
Germfree	2.7 ± 0.49[3]	0.5
Conventionalized[2]	3.2 ± 0.59	0.5
E. Coli[2]	3.0 ± 0.32	0.7
Bacteroides[2]	3.1 ± 0.54	....

[1] Pooled samples.
[2] Rats monocontaminated or conventionalized for 2 weeks.
[3] Mean ± SE.

tinal mucosal epithelial cells might account in part for changes in whole body metabolic rate of rats with different intestinal bacteria. This occurred to us because of observation of Abrams, Bauer, and Sprinz (61) of a lower turnover rate of ileal mucosal epithelial cells in germfree mice than in conventional mice.

This hypothesis was tested with littermate male Fischer rats, one group of which was germfree, one monocontaminated with *E. coli,* one with *Bacteroides,* and one group conventionalized. Purposeful contaminations were made at weaning. Two weeks later O_2 consumption and CO_2 production increases induced by *E. coli* contamination and conventionalization had peaked and stabilized. H^3 thymidine was then injected intraperitoneally and rats of each group were killed 12, 36, and 60 hours later. Migration of thymidine tagged mucosal epithelial cells was determined by an autoradiographic histologic method. Data appear in Table 3-VIII.

TABLE 3-VIII

Turnover Rate of Ileal Mucosal Epithelial Cells of Fischer Rats* Migration of Lead Cell as Percent Villus Length

Time[+]	*Conventionalized* (5)	*E. Coli* (5)	*Germfree* (5)	*Bacteroides* (5)
12 hrs.	34.7 ± 1.82	36.3 ± 1.14	35.1 ± 1.66	34.0 ± 2.18
36 hrs.	66.5 ± 2.36	69.9 ± 3.04	59.2**	59.2 ± 1.97
60 hrs.	93.0 ± 2.02	90.3 ± 2.46	88.1**	86.3 ± 2.13
Mitoses***	196	....	203	....

[+] Time after i.p. injection of H^3-thymidine.
* Male rats, about 3 months old.
( ) No. of rats.
** Not completed.
*** No. cells in mitoses per 100 crypts.

Turnover of ileal mucosal epithelial cells of conventionalized and *E. coli* mice was very similar and was only slightly higher than turnover rates of GF and Bacteroides rats. This difference, however, was very slight, three to seven percent, far from the 100 percent difference reported by Abrams, *et al.* (61) between germfree and conventional mice. Supporting this is the fact that mitotic rates of ileal mucosal cells were similar in all groups in our experiment.

The dramatic effect of the infectious state which characterized normal healthy mammalian life on resting metabolism has not been realized until recently. No mention of this is made in any of the standard books or discussions of basal metabolism. Effects of infectious diseases, of course, have been long known. In addition to our observations, Desplaces, Zagury, and Sacquet (62) have found that oxygen consumption of open animal room rats was higher than that of germfree rats. Windmueller, McDaniel, and Spaeth (63) have reported that carbon dioxide production by open-animal room rats is higher than that by germfree rats. Wostmann, Bruckner-Kardoss, and Knight (64) have also reported that germfree rats have lower oxygen consumption rates than open animal room or conventionalized rats.

Our experiments indicate that faculative anaerobes, such as E. coli, and not strict anaerobes such as Bacteroides sp. increase oxygen consumption and carbon dioxide production by rats, directly or indirectly. These changes were independent of sex or colonic temperature change in the E. coli experiments.

We base this theory on the fact that changes in oxygen consumption and carbon dioxide production are prompt, beginning within hours after purposeful contamination with *E. coli.* They do not follow contamination with *Bacteriodes.* We think that an important part of the effect is direct utilization of oxygen and production of carbon dioxide by facultative anaerobic bacteria. Windmueller and his associates also suggested that these factors may be involved. This, however, remains to be proved.

Some information is available about rate of oxygen consumption by *E. coli* growing in culture with excess oxygen present, but not under conditions similar to those present in the gut. Under the former condition rate of oxygen consumption is very high.

Indirect effects on the host are also suggested. We found a lag, at times of a few days, between peak rise in oxygen consumption and carbon dioxide production and number of viable bacteria in feces of rats purposefully contaminated with *E. coli.* These indirect effects would include possible alteration of thyroid function as suggested by Desplaces and her associates. (62) Their data are not convincing to us, however. As mentioned, our own data show no differences in serum PBI and serum thyroxine iodine concentrations of germfree and conventionalized rats.

We plan to explore formation of the amine, tryptamine, by gut bacteria as a possible mechanism by which bacteria influence resting metabolism of the animal host. This thought occurred to us because of Allen's (65) demonstration that such amines may increase oxygen consumption of chicks. Wostmann, Bruckner-Kardoss, and Knight (64) has postulated that absorption of one or more bioactive substances from enlarged ceca of germfree rats might, by some unknown mechanism, exert a depressant action of oxygen consumption and cardiac output. In this regard, Gordon (66) has described toxic and musculoactive principles present in greater amounts in cecal contents of germfree animals.

The reason for higher colonic temperature of conventionalized rats as compared with germfree and possible effects of this colonic temperature differential on responses of rats to a variety of challenges are to be studied. Factors underlying differences in colonic temperatures of germfree and conventionalized rats are probably not the same as those underlying differences in oxygen consumption and carbon dioxide production. The former, nevertheless, may include the latter. In this regard, recall the dichotomy between O_2 consumption, CO_2 production and body temperature of rats monocontaminated with *E. coli* in comparison with their littermates which remained germfree.

Response To Starvation

In view of the lower metabolic rate of germfree rats, it was surprising to us that they die faster than conventional and conventionalized rats when starved. Similarly, germfree mice die faster than conventional and conventionalized mice (Fig. (3-5). (67,68,69)

We conducted an experiment to see whether changes in metabolic rates during starvation might account for differences in survival time between germfree and conventionalized rats. (68) During starvation, there were prompt and parallel drops in O_2 consumption in both germfree and conventionalized rats. By the tenth day, these values were 20 percent lower than prestarvation control values. Oxygen

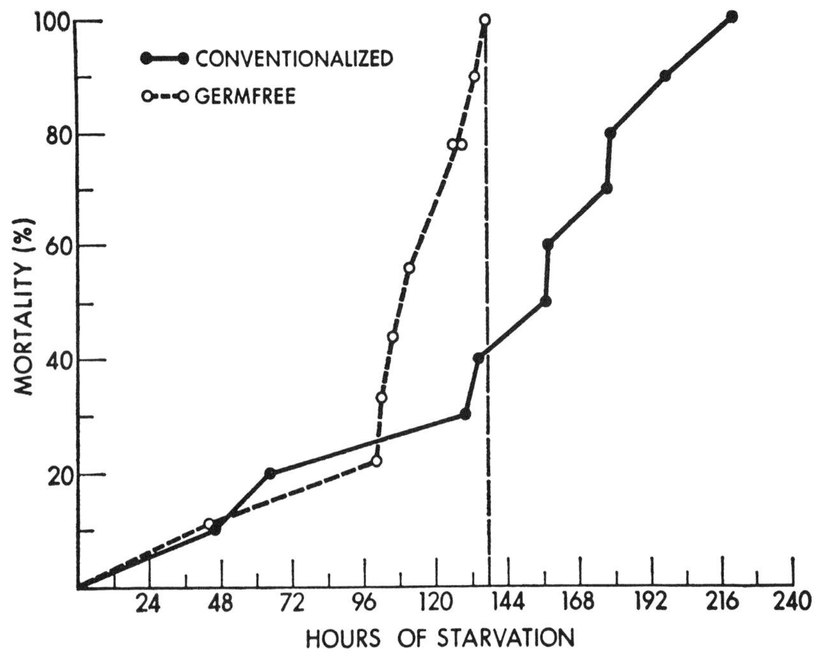

Figure 3-5. Starvation of germfree and conventionalized mice: cumulative mortality. (Reproduced through the courtesy of *J Nutr, 94:* 151, 1968.)

consumption of conventionalized rats continued to drop at the same rate. It was about 30 percent lower on the 13th day and 37 percent on the 18th day. In contrast, oxygen consumption of germfree rats fell sharply between the 10th and 13th days. It was about 60 percent lower than control values on the morning of the 13th day. All the germfree rats were dead by the afternoon of the 13th day, while conventionalized rats died between the 18th and 21st days of starvation. The drop in O_2 consumption fell faster than body weight in conventionalized rats until the 10th day of starvation. Following this, change in oxygen consumption and body weight were similar until the 18th day, the last day of measurement. In germfree animals, O_2 consumption also fell faster than body weight until the 13th day. On that day, oxygen consumption dropped precipitously while decline in the weight continued unchanged. Carbon dioxide production fell faster than the oxygen consumption in both groups. As a result, the R. Q.'s fell from 0.84 to 0.79 by the third day. The R.Q. remained at this level until the 13th day when it fell still further in both groups. Colonic temperature in the prestarvation control period averaged 37.4°C in the germfree, and 37.8°C in the conventionalized rats, a difference that we have noted previously. In each group, colonic temperature dropped 0.4 to 1.0°C in the first three days and remained at these levels until about 12 hours before death when they fell sharply. These falls occurred later in conventionalized animals because they lived much longer than germfree rats (Fig. 3-6).

These data demonstrate no proportionate relative "increase" in energy metabolism of germfree rats during the first 10 days of starvation to account for their earlier death. On the other hand, energy production and colonic temperature fell abruptly on the 13th day, 12 to 24 hours before death.

Starved germfree rats and mice lost weight at the same rates as their conventional and conventionalized counterparts. Even though germfree animals died having lost substantially less weight (Fig. 3-7), rates of loss of certain body

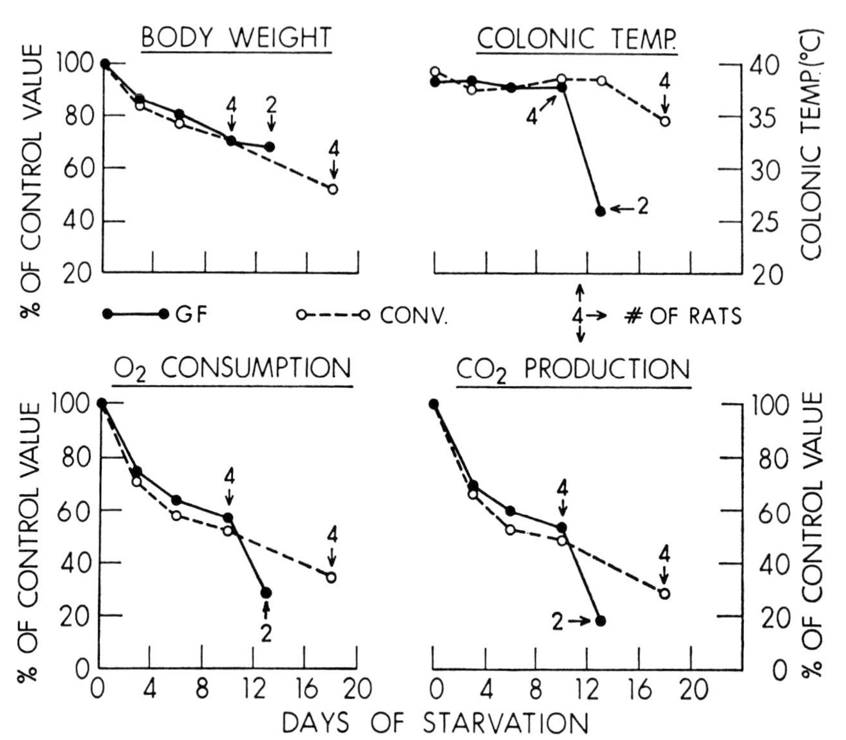

Figure 3-6. Changes in oxygen consumption, CO_2 production, colonic temperature and body weight of germfree and conventionalized rats during starvation.

constituents differed. For example, in one experiment, the germfree rats weighed about 10 percent more than the conventionalized rats (364 g and 330 g, respectively, $p < 0.02$). This was due to greater weight of gastrointestinal tract contents of germfree rats, 61.5 g vs. 18.3 g, $p < 0.001$. Carcass weight of germfree rats was less than that of conventionalized, 252 g vs. 270 g, $p < 0.02$. The empty gastrointestinal tract of the germfree was slightly heavier, 10.7 g vs. 8.9 g, $p < 0.01$. Weight of liver, heart, kidneys, lungs, spleen, and adrenals were similar in both groups.

As for the chemical composition, percent protein in carcass was the same in both groups. Percent fat was significantly

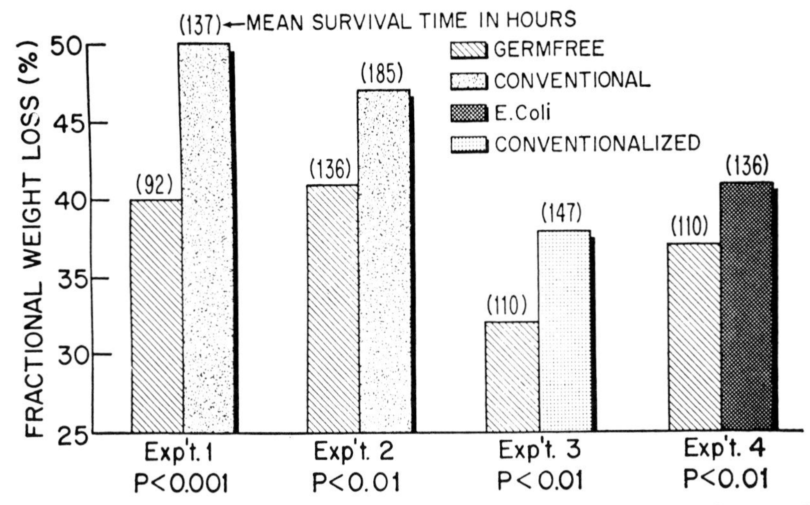

Figure 3-7. Relationship between survival during starvation and fractional weight loss of germfree, conventional, *E. coli* monocontaminated, and conventionalized mice. (Reproduced through the courtesy of *J. Nutr, 94:* 151, 1968).

less in the germfree, 13.2 percent vs. 18.5 percent, $p < 0.05$. Percent water was significantly higher in the germfree, 59.7 percent vs. 56.0 percent, $p < 0.01$. Similar differences in percent fat and water were found in the gastrointestinal tracts. In contrast, liver chemical composition was similar in both groups.

Following starvation, rats were killed at 11 days, a time one to three days before germfree rats of this age generally die. Both groups had lost about one-third of their body weight, 34.2 percent and 33.6 percent, a difference not significant (NS), and carcass weight, 29.1 percent and 29.4 percent, NS. Empty gastrointestinal tracts dropped similarly in weight in both groups. Conventionalized animals, however, lost a greater proportion of their gastrointestinal tract contents, 78 percent vs. 47 percent, $p < 0.001$. The proportions of carcass fat, protein and water lost were different. Thus, carcass fat fell almost 75 percent in the germ-free, and 60 percent in the conventionalized rats ($p < 0.05$).

During 11 days of starvation, germfree rats lost significantly more nitrogen in their urine than did conventionalized, 870 mg vs. 675 mg, $p < 0.05$. Fecal nitrogen was similar in both groups. In fact, very little fecal nitrogen was lost after the first dew days. Urinary sodium, potassium, and chloride excretions were similar in the two groups.

In starved mice, carcass fat also fell significantly faster in germfree, averaging about 80 percent in the germfree and about 45 percent in the conventionalized after 96 hours, $p < 0.001$. Liver fat rose strikingly higher in the germfree mice by 48 hours, about 140 percent vs. 50 percent, $p < 0.001$, and was proportionately higher at 72 hours. By 96 hours, liver fat was still increased about 25 percent in the germfree mice, while it was decreased about 25 percent in the conventionalized (Figs. 3-8, 3-9).

In the mice experiments, the most striking blood chemical finding was that blood sugar concentrations dropped significantly faster in the germfree and reached lethal levels.

To summarize, data of rat and mice experiments during starvation indicate the following. A greater mobilization of carcass fat occurs along with greater accumulation of fat in liver early, and lesser ability to maintain blood sugar in the germfree animals as compared with conventionals. It will be recalled that normal, unstarved germfree rats have less carcass fat than their conventional counterparts. Mechanisms underlying these metabolic differences are not known.

We did an experiment to test the hypothesis that conventional rats survived longer by getting nutrients from bacterial action by coprophagy. Such nutrients would not be available to the germfree. Our data do not support this hypothesis.

Amino Acid and Protein Metabolism

Earlier salutary effects of certain oral antibiotics on growth (70–72) demonstrated profound effects of indigenous microflora, especially gut bacteria on amino acid and protein metabolism of the animal host. This antibiotic action has profoundly affected meat and poultry industries and world-

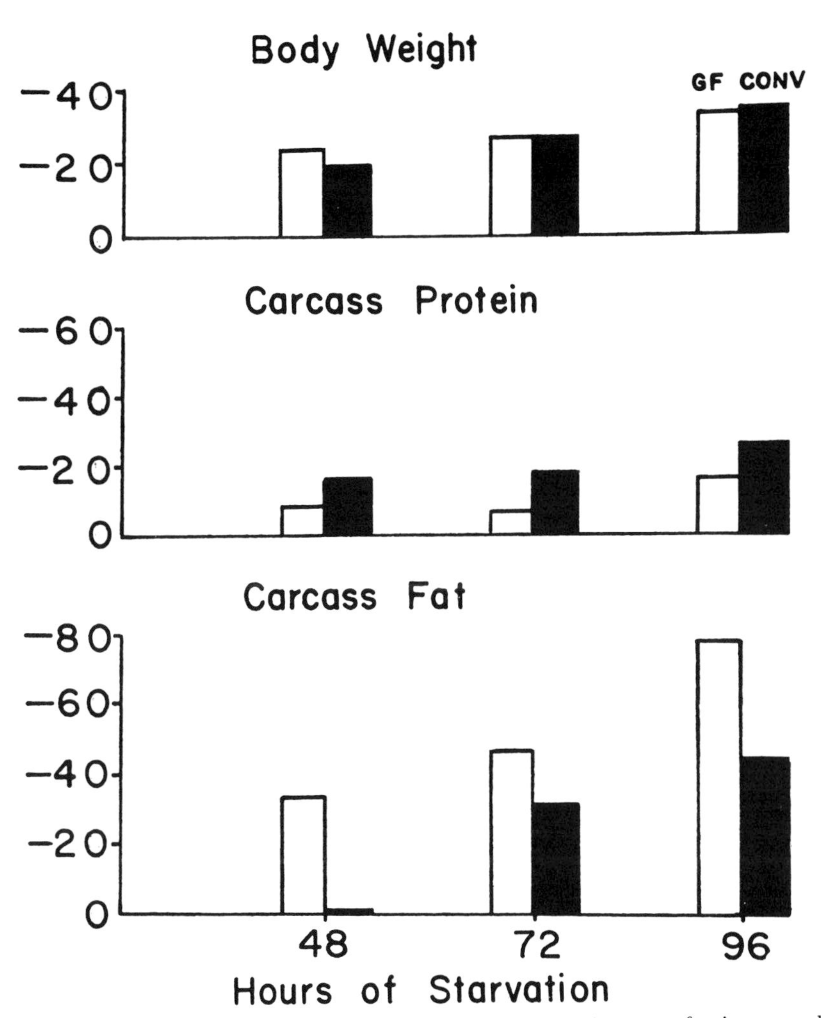

Figure 3-8. Changes in body weight, carcass protein and carcass fat in starved germfree and conventionalized mice.

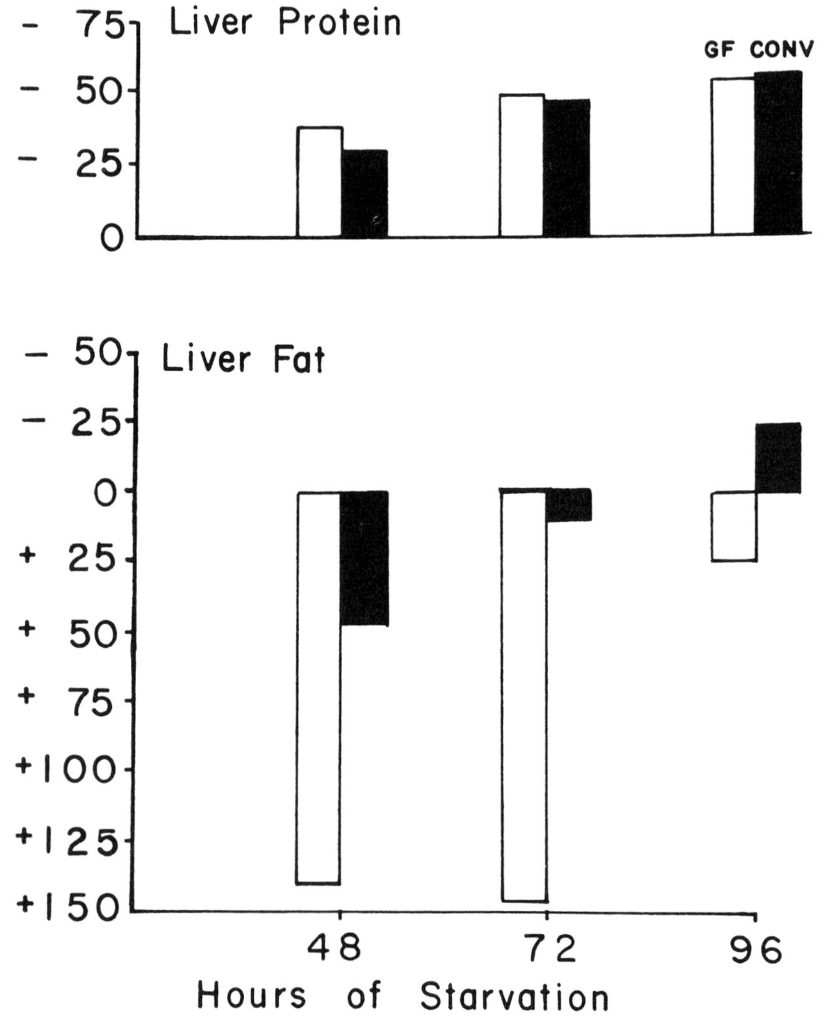

Figure 3-9. Changes in liver protein and liver fat in starved germfree and conventionalized mice.

wide economy. With the advent of the modern era of germfree technology, some studies were directed towards elucidating mechanisms underlying antibiotic effects on improved feed efficiency and growth of conventional animals. Antibiotic effects are clearly due to antibacterial effects, resulting in a change in the gut flora. (73,74) Data thus support Mechnikoff's view. (6) Half a century or more earlier he pointed out that certain gut bacteria are detrimental and some beneficial to health of the host animal, including humans.

Degradation of certain amino acids by gut bacteria has been known for a long time. Examples are:

Lysine ⟶ Cadaverine
Arginine ⟶ Ornithine
Histidine ⟶ Histamine
Tyrosine ⟶ Tyramine
Tryptophan ⟶ Indole and Tryptamine

Michel (75) reported that all L-amino acids may be degraded by gut bacteria. Reactions involved are principally deaminations and decarboxylations. Ammonia, carbon dioxide, hydrogen sulfide, amines, and various mercaptans are formed. Certain gut bacteria enzymatically hydrolyze urea in mammals (76,77,78) Fig. 3-10). Significance of some of these reactions, especially for patients with hepatic disease, and portal hypertension has been known for a long time. Protein restriction, antibiotic administration, and feeding of lactobacilli to change gut flora (79–84) have been tried to decrease these reactions.

Doft and McDaniel (85) conducted studies with weanling rats to determine quantitative significance of degradation of certain amino acids by gut microflora. In early experiments, they noted that germfree rats of the Lobund strain given a low-protein liver necrogenic diet often grew more rapidly than did conventional rats on the same diet. To determine whether this finding depended primarily on diet employed, the most limiting amino acids of this diet, threonine, tryptophane and methionine, were added to the basal diet and

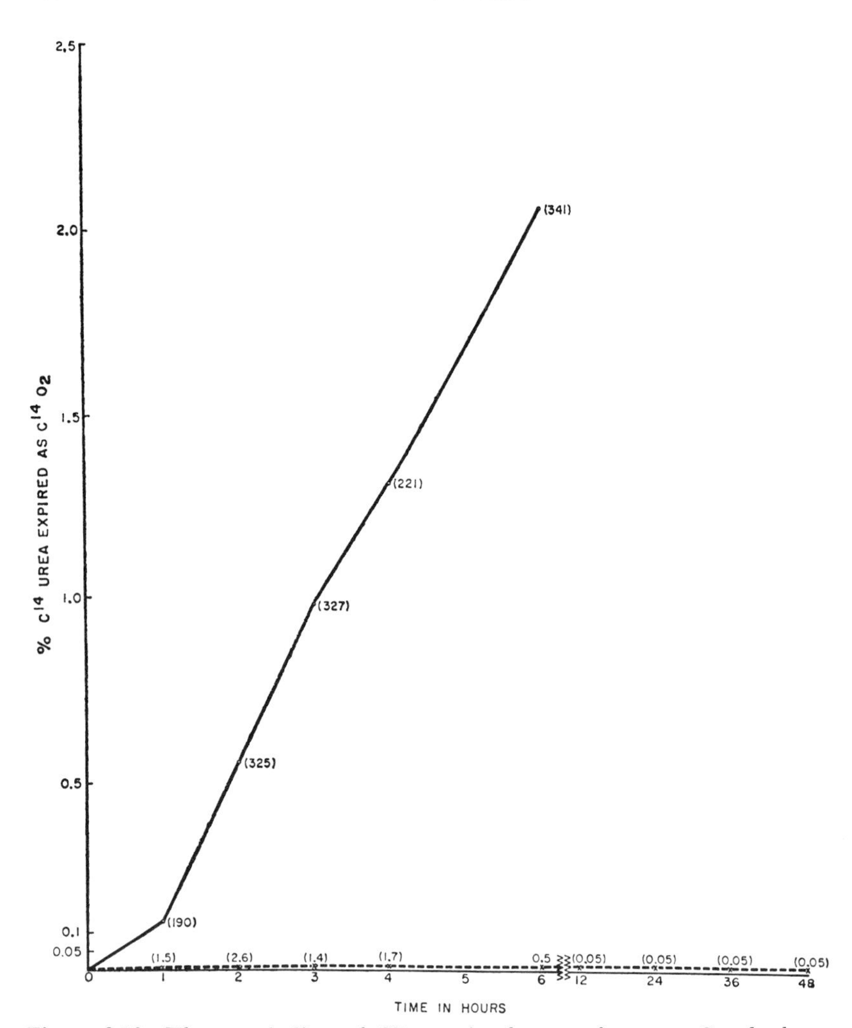

Figure 3-10. The metabolism of C^{14}-urea in the germfree rat. Graph shows the percentage of the administered C^{14} expired as $C^{14}O_2$ and the specific activities of the CO_2 expired. Solid line represents the conventional rat; the dotted line the germfree rat. (Reproduced through the courtesy of *J Biol Chem, 234:* 2061, 1959).

fed to germfree and conventional rats. Growth rate of conventional but not of germfree animals was increased; germfree and conventional rats now grew at essentially the same rate.

In view of these data, diets deficient in each of the ten essential amino acids were studied by them. The rats were fed the diets *ad libitum* starting at weaning. Germfree rats appeared to grow much better than their conventional counterparts when the diet was deficient in tryptophan, arginine or lysine but not on diets deficient in valine or containing adequate amounts of all the ten essential amino acids. A small growth differential, in favor of the germfree rats, was noted when diets deficient in one of the other essential amino acids were employed.

A probable explanation for greater growth rates of germfree rats on diets limited in certain amino acids is that bacteria of gastrointestinal flora compete with host animals for amino acids. Certain ones such as tryptophan, arginine, and methionine are metabolized competitively to a greater extent than others, even when in short supply in the diet and badly needed by the host animal. A rough estimate of possible net destruction or utilization of specific amino acids caused directly or indirectly by microflora of conventional rats was calculated. This was done by determining amount of a given amino acid which had to be added to the 4 percent casein basal diet to bring growth rate of conventional rats to that attained by germfree. These values were equivalent to about 40 percent of the methionine, 45 percent of the threonine, and 50 percent of the tryptophan in the basal diet.

Diets used in the above experiments contained adequate amounts of the vitamins needed by the rat, including niacin. Further experiments have been carried out by Doft and McDaniel to study effect of the germfree state on rats receiving diets deficient in both tryptophan and niacin. The classical Goldberger diet (109) with vitamins other than niacin added and with tryptophan was used. It permitted the conventional rat to grow at a subnormal rate with no evidence of "blacktongue" or similar pathological conditions. Germfree rats on this diet grew considerably more rapidly than did their conventional counterparts. Consumption of a diet deficient, as the Goldberger diet, in niacin but also critically deficient in tryptophan led to the consistent and early death of weanling conventional rats. Eight percent casein, 5 percent zein and 5 percent gelatin served as the source of protein. Germfree rats, on the other hand, survived many months even though they grew slowly. It appears, therefore, that in the rat, at least part of the deleterious effects of niacin-tryptophan deficient diets depends on presence of bacteria. These bacteria are most likely gastrointestinal. Presumably they compete with the host for trpytophan and thus bring about a more acute niacin deficiency.

In a related type of experiment, Seifter, Rettura, and Levenson (86) determined the effect of methionine on the growth of conventional and germfree rats eating the Salmon-Newberne choline-deficient diet. (110) We found that open animal room rats respond, in terms of growth, better to choline than to methionine. In an experiment with germfree rats on this diet, growth response to methionine was considerably better than to choline. The following theoretic explanation is offered. By analogy to bacterial decomposition of choline to trimethylamine by intestinal flora, methyl mercaptan, the disulfide or dimethyl sulfide are probably produced from methionine by bacteria and these compounds may have growth depressing effects. The reactions probably do not occur in germfree animals.

Conversely, evidence indicates that gut bacteria may synthesize certain amino acids which may be available to the host for its general nutrition in a limited way. For example, Seifter *et al.* in our laboratory (87) studied growth response of rats to shikimic acid. This is a known precursor of phenylalanine biosynthesis in bacteria. Two compounds, quinic acid and phenyllactic acid, chemically related to other intermediate compounds of phenylalanine biosynthesis, were also studied. It was expected that these compounds would stimulate phenylalanine biosynthesis by gut flora. By either coprophagy or direct absorption, such phenylalanine would positively influence growth when rats were grown on a phenylalanine-deficient diet. Young male rats of the Fischer strain were fed either purified or chemically defined low phenylalanine diets for periods of 5 to 10 days. L-Phenylalanine supplementation permitted a growth response of 4 g/day with a feed efficiency of 3. Feed efficiency is defined as the ratio of grams of feed consumed to grams of weight gained. L-Phenyllactic acid was not as effective as phenylalanine. It produced maximal growth rate approximately one-half of that induced by phenylalanine. Activity of shikimic acid was even lower. It was equivalent to one-quarter of the rate with phenyllactic acid. Shikimic acid feeding produced a response similar to that of feeding phenylalanine at a dietary level of 0.04 percent, corresponding approximately to 10 percent of the phenylalanine requirement of the rat. Feeding D-quinic acid was without effect. When rats were treated with neomycin sulfate, the growth response to shikimic acid was not seen. The data suggest that quinic acid cannot be converted to phenylalanine by the rat or its intestinal flora, that limited amounts of shikimic acid are converted to phenylalanine, and that this conversion requires an intestinal flora.

We will now discuss *toxicity of amino acids, their metabolites and amino acid esters* for rats, and how toxicity is affected by indigenous microflora. Harper and his associates have published recently a very extensive reviw of amino acid imbalances, antagonisms, and toxicities. (88) We became interested in this problem serendipitously. Several years ago we began some nutritional studies in germfree and conventional rats using chemically-defined liquid diets of the type described by Greenstein, Winitz and their associates. (89–100) Such diets are made up of L-amino acids, the simple sugars, glucose or sucrose, minerals, known vitamins and accessory food factors, a synthetic unsaturated fatty acid, ethyl linoleate, water, and, at times, an emulsifying agent, polysorbate. We called them J_2 and began using this type of diet because routine diets for germfree animals, sterile from steam autoclaving or irradiation, contain dead bacteria and other protein and carbohydrate polymer antigens. These diets complicate immunologic reactions of germfree animals and interfere with study of "natural" resistance and fundamental interactions among animals, microorganisms, and allergens. Chemically-defined diets of highly purified recrystallized low molecular weight compounds avoid these complications. Special advantages are more precise investigations of specific metabolic and nutritional needs of experimental animals as well as effect of specific microbes on mammalian metabolism and nutrition.

During these studies, we came across two totally unexpected findings in ordinary rats ingesting this type of diet. In a few weeks they developed (1) pancreatic acinar atrophy and (2) hemolytic anemia. No lesions, or much less severe lesions occurred in germfree rats. (101–104)

Microscopically, the *pancreatic lesion* was characterized by progressive quiet necrosis of acinar cells followed soon by fibrosis. Acinar atrophy and fatty replacement was seen after two or three months (Fig. 3-11). There was no glycosuria and the islets were not involved for many months; peri-islet metaplasia occurred after 6 or more months, but origin of metaplastic cells is not known. (102) Pancreatic enzymes were not measured directly, but fat absorption was greatly impaired as measured by a test dose of I^{125} triolein. Digestion and absorption of proteins or carbohydrate polymers have not been tested yet.

Livers of rats are normal grossly and histologically (light microscopy) as are *kidneys*, though *azotemia* develops in conventional rats.

In this same series of experiments, severe anemia developed, almost invariably (99%) in conventional rats, but in less than 1 percent of germfree rats. Anemia comes on rapidly, hemoglobin dropping to 6 to 10 gm percent in 3 to 4 weeks and often falling as low as 2 to 3 gm percent (Fig. 3-12). Heart, spleen, and kidney enlarge and rats often die of congestive failure. Anemia and cardiac enlargement also develops in ordinary rats maintained in isolators, so-called IHO rats. The same is true for germfree rats which are purposely contaminated with the cecal contents of ordinary rats, so-called conventionalized rats, but not their germfree litter mates.

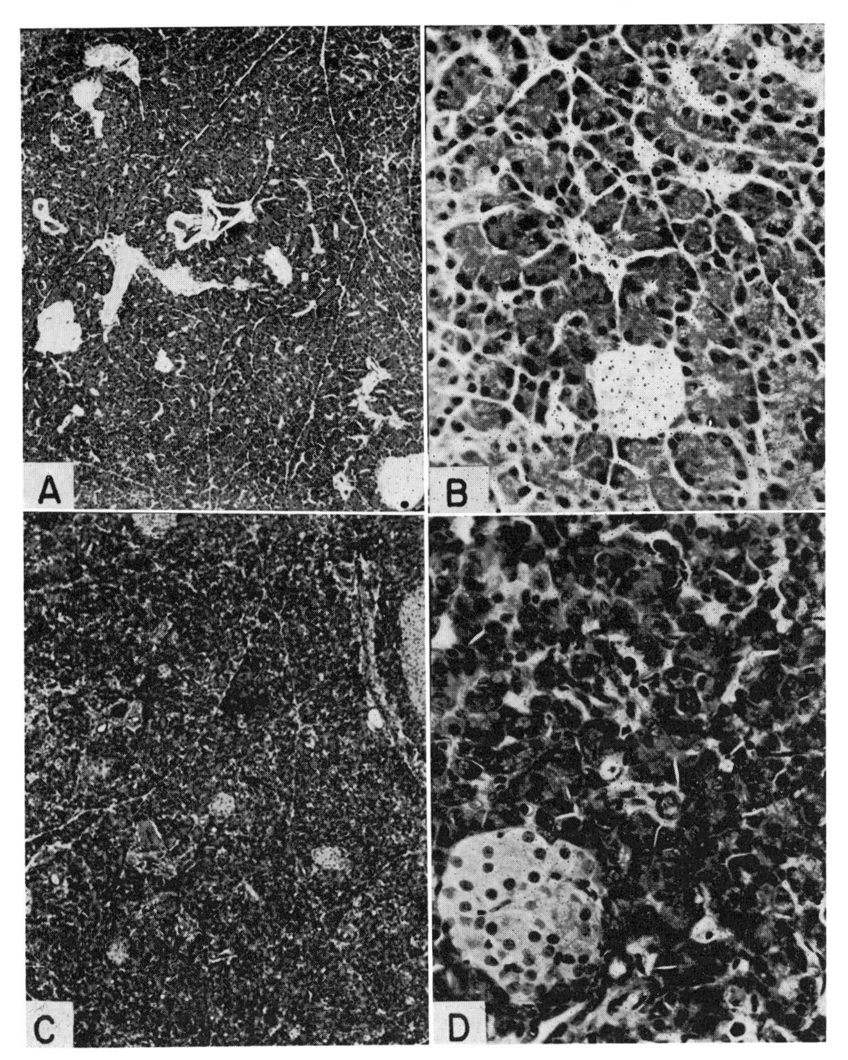

Figures 3-11 A,B,C,D. Pancreatic lesion (H. and E. stain; A, C: × 60; B, C: × 400).
(A) (B) Grade 0: normal. Regular acinolobular organization.
(C) (D) Grade 1: loss of regular acinolobular pattern, irregular nuclei, and cytoplasmic vacuolization.
Pancreatic lesion (H. and E. stain; E, G: × 60; F. H: × 400).
(Reproduced through courtesy of Ann Surg, 174: 469–510, 1971.)

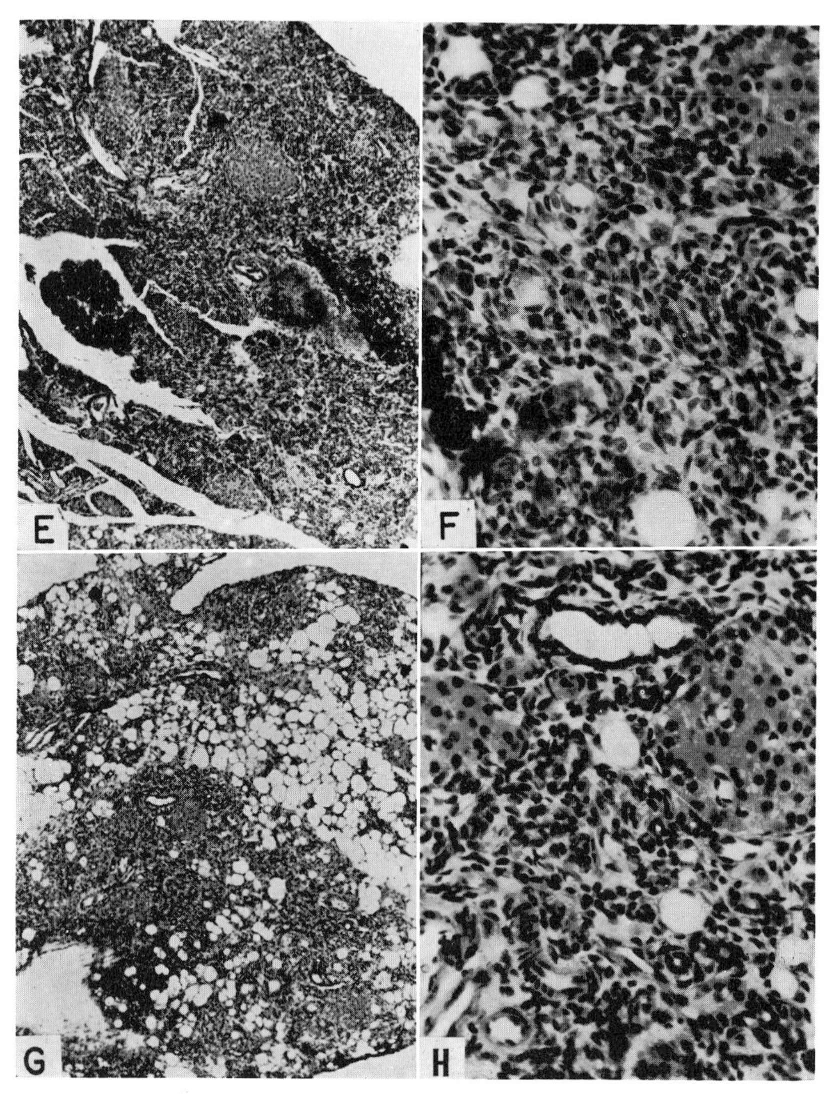

Figures 3-11 E,F,G,H. Grade 2: no regular acini, marked decrease of the acinar cells, and marked fibroblastic proliferation.
(G) (H) Grade 3: diffuse fibrosis, very few acinar cells, fatty metamorphosis, and prominent islets. (Reproduced through the courtesy of *Ann Surg, 174:* 469–510, 1971.)

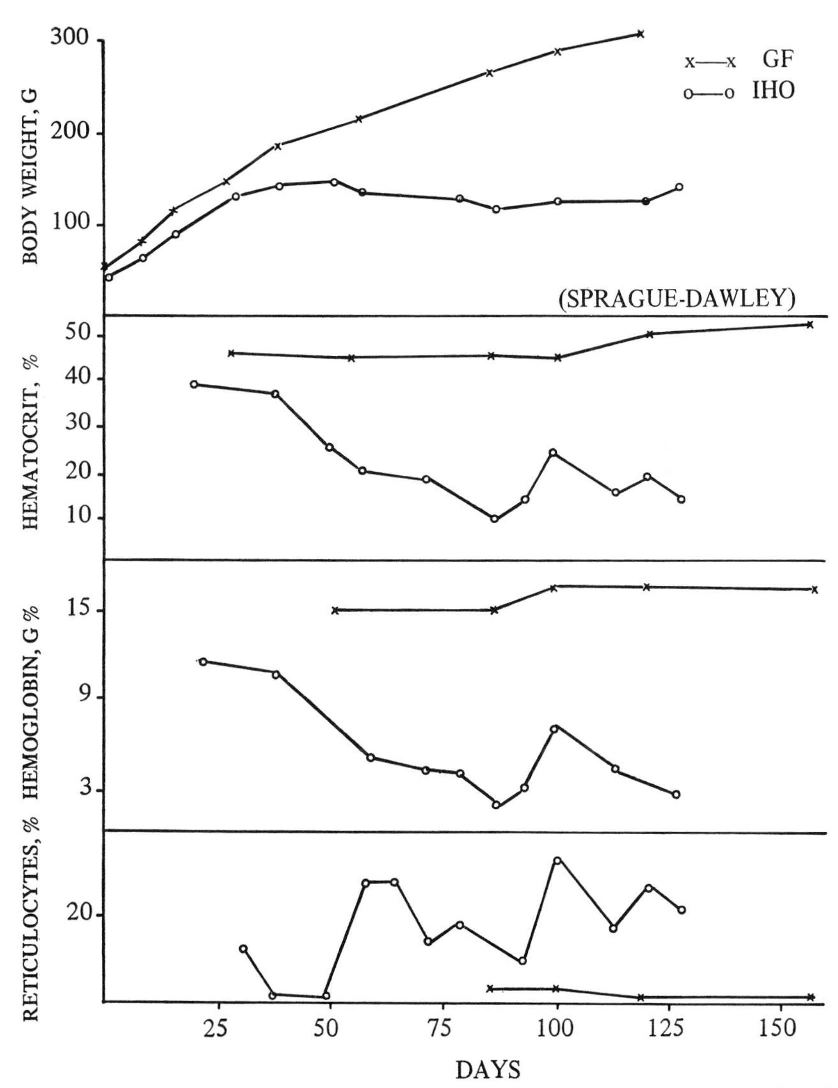

Figure 3-12. Typical GF and IHO rats fed liquid diet J_2. (Reproduced through the courtesy of the *Ann Surg, 174:* 469–510, 1971).

Anemic rats have no blood loss. Serum iron levels and serum total iron binding capacities are not changed. Bone marrow shows normoblastic erythroid hyperplasia and peripheral reticulocytosis occurs. The spleen is enlarged and has greater than normal amounts of iron.

Red blood cell incorporation of intravenously injected radioactive iron is more rapid in anemic rats, a finding expected in view of the marked reticulocytosis. Very rapid radioactivity decline also occurs in anemic rats. Both are in contrast to germfree rats. These data indicate that anemia is hemolytic. In fact, red blood cell survival is decreased by a factor of 2, as indicated by the fate of transfused 51 chromium-labeled autogenous cells (Fig. 3-13). Cross-transfusion experiments with labeled red blood cells demonstrated that the "defect" leading to the hemolytic anemia resides in the red blood cell. Figure 3-14 demonstrates that red blood cells of normal rats eating chow or germfree rats ingesting liquid diet J_2 survive normally when auto-transfused or cross-transfused into other normal or anemic rats. In contrast, red blood cells from anemic rats ingesting liquid diet J_2 show the same shortened survival times when auto-transfused or cross-transfused into other anemic or normal rats.

Splenectomy lessened but did not cure or prevent anemia.

In collaboration with Drs. Jaffe and Hsieh, we conducted experiments to determine the nature of the red blood cell defect. The most striking change is an abnormal hemoglobin absorption spectrum pattern, characteristic of "sulfhemoglobin." The abnormal spectrum was found in 14 of 15 anemic rats and none of 22 control rats.

There has been no evidence for a vitamin deficiency and no single vitamin or combination of vitamins is effective in treating or preventing the anemia.

Anemia and azotemia could be corrected or prevented by solid supplement of proteins from diverse sources, as ultrapure bovine serum albumin, ultrapure α-lactalbumin or vitamin-free casein. Vitamin-free casein hydrolysate, an acid hydrolysis - tryptophan added, was also effective, but a mixture of free amino acids in equivalent and proportionate amounts as the casein supplement was not. Further, supplements of casein or casein hydrolysate ameliorate but do not wholly prevent the pancreatic lesion. Supplement of amino acids in amounts and proportions similar to casein does not (Table 3-IX) (Fig. 3-15). Ashed casein had no prophylactic

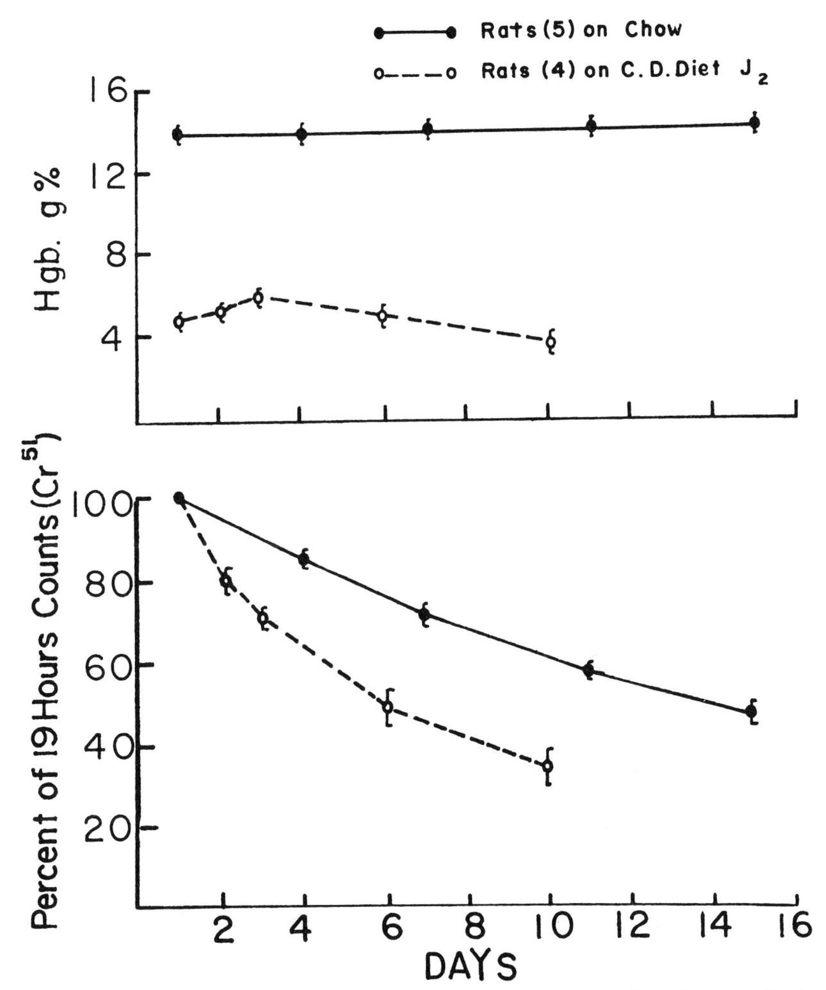

Figure 3-13. Red blood cell survival of S-D rats. (Reproduced through the courtesy of *Ann Surg, 174:* 469–510, 1971.)

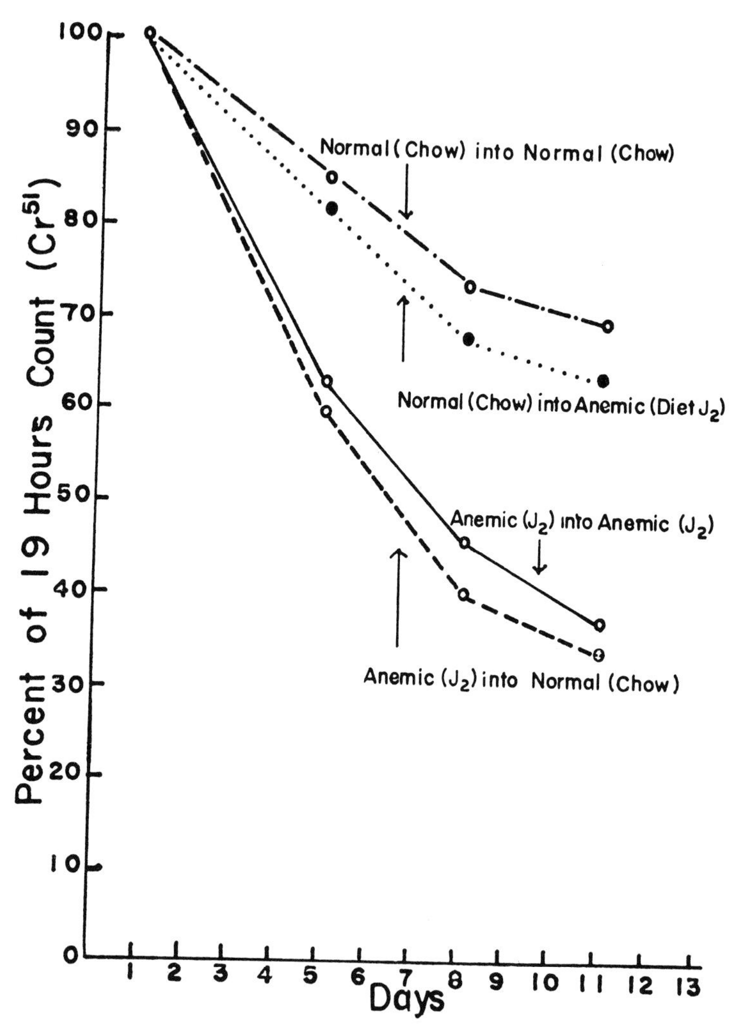

Figure 3-14. Red blood cell survival of S-D rats after transfusion. (Reproduced through the courtesy of *Ann Surg, 174:* 469–510, 1971.)

benefits when fed as a supplement. Beneficial effects of casein thus are not due to some heat stable trace component, such as trace metals. These data suggest a possible special hitherto unrecognized role for dietary peptides or nucleotides in mammalian metabolism.

TABLE 3-IX

Effects of Various Prophylactic Supplements on the Development of the Anemia and the Pancreatic Lesion by Rats Ingesting Liquid Diet J_2; Effect of Solid Diet J_2

	Anemia	*Pancreas*
Ineffective	Glucose Amino Acids Ashed Casein	Glucose Amino Acids Ashed Casein
Moderately Effective		Casein Casein Hydrolysate Solid Diet J_2
Very Effective	Casein Casein Hydrolysate Solid Diet J_2	

(Reproduced through courtesy of *Fed. Proceedings, 30:* 1785–1802, 1971.)

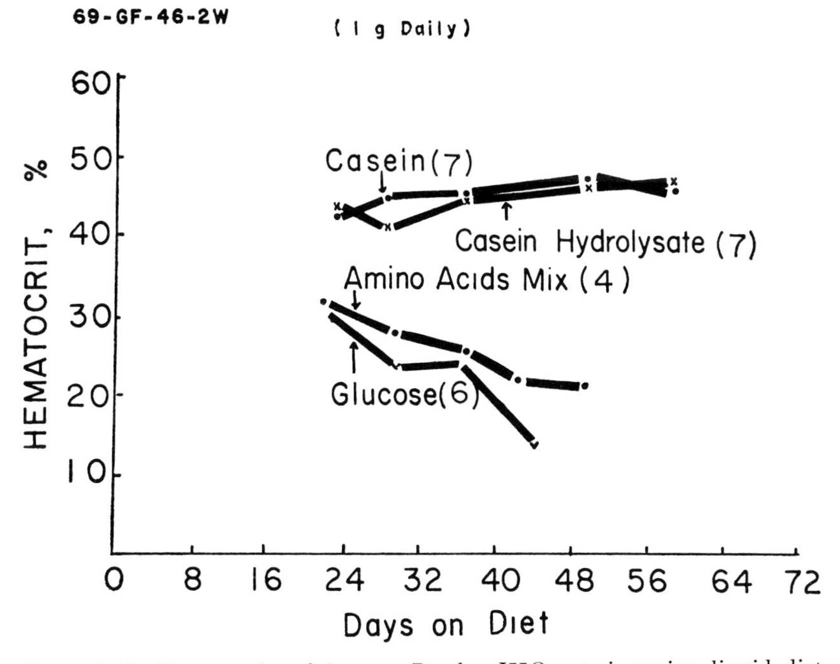

Figure 3-15. Hematocrits of Sprauge-Dawley IHO rats ingesting liquid diet J_2 and given various oral supplements prophylactically. (Reproduced through the courtesy of *Ann Surg, 174:* 469–510, 1971.)

Development of anemia, azotemia, and pancreatic lesions depend on some coincidence of dietary ingredients and rat microbial flora. The "chemically defined" amino acids are most toxic when prepared and ingested in liquid form.

Critical microbial factor (s) is present in whole cecal contents, but not ultrafiltrate of cecal contents, of healthy conventional rats eating chow. This was evident with germfree rats purposefully contaminated with cecal contents of conventional rats eating a commerical rat chow "complete" diet. These animals then became anemic, azotemic and developed pancreatic lesions when they were switched to the liquid amino acid diet J_2. In contrast, their germfree littermates which were kept germfree did not, despite ingestion of the same liquid diet. Other experiments indicated that gut bacteria were almost certainly the critical component. In particular, *Proteus sp.*, and possibly *Staphylococci,* are involved.

The syndrome in rats ingesting liquid diet J_2 differs substantially from abnormalities described by other investigators. They have been studying (a) "chemically defined" amino acid diets; (b) pancreatic acinar atrophy and fibrosis; (c) nutritional or other hemolytic anemias and (d) amino acid imbalances, antagonisms and toxicities including those of methionine, homocyst (e) ine, cystine, and cysteine. A review of this is published (102–104) and a detailed paper is in preparation.

Amino acid formulation of three liquid diets, 116 and J_2 of Greenstein, Winitz, and Otey (100) and L-479E of Reddy, Pleasants and Wostmann (105), are listed in Table 3-X. Diet J_2 leads to the pathologic syndrome of anemia, azotemia, and the pancreatic lesion. Rats ingesting diet 116 do not become anemic or azotemic. They often develop the pancreatic lesion, though to a less extent than rats ingesting diet J_2. (104) Rats ingesting diet L-479E do not develop the pancreatic lesion. (105)

The higher concentration of cysteine ethyl ester in diet J_2 was suggested as critically involved first by E. S. of our group. This was because we (106) and Shapiro *et al.* (107)

TABLE 3-X

Composition of Liquid Amino Acid Diets*

	116	*J_2*	*L-479E*
Amino acids		*g/liter Diet*	
L-Cysteine ethyl ester HCl	0.55	2.42	
L-Tyrosine ethyl ester HCI	8.40	3.14	10.00
L-Asparagine		3.95	6.00
L-Serine	6.60	2.30	7.75
L-Proline	12.70	2.30	15.00
L-Aspartate	6.75	2.30	
L-Glutamate, mono-Na	25.60	27.95	30.00
L-Alanine	3.20	2.30	3.75
L-Arginine HCl	4.70	8.90	3.75
Glycine	2.05	13.94	2.50
L-Histidine HCl H_2O	2.85	3.38	2.75
L-Isoleucine	4.40	5.50	2.50
L-Leucine	7.00	7.30	4.00
L-Lysine HCl	6.50	11.80	6.25
L-Methionine	3.15	5.39	4.25
L-Phenylalanine	3.15	7.63	4.50
L-Threonine	4.40	5.00	2.50
L-Tryptophan	1.40	1.50	2.00
L-Valine	4.90	5.50	3.50
	108.30	122.50	111.00

* A table showing compositions of the complete diets (including vitamins, minerals and fatty acids) is in *Federation Proceedings, 30:* 1785, 1971 (103).

had previously found that it combined with menadione to form an adduct which interfered with prothrombin synthesis and led to vitamin K deficiency. This does not happen when vitamin K_1 is used as in diet J_2. In addition, we could visualize it interacting with other ingredients in the diet and/or components in the rat's gut, including bacteria and their products, or metabolites resulting from interaction of bacteria and other compounds. As a result, compounds may be formed which might do the following. They could give rise to H_2S or acrylyl acid derivatives. They might effect condensation of acrylyl or pyruvoyl derivatives to various amines, or compounds which are amino acid antagonists as lysine analogs.

Experiments tested the possibility that cysteine ethyl ester hydrochloride (C.E.E.HCl) is involved in pathogenesis of the syndrome. Diet J_2 was prepared with and without cysteine ethyl ester. Normally, C.E.E.HCl is the last solid ingredient added in preparation of diet J_2. A batch of diet was prepared and halved. C.E.E.HCl in normal concentration was added to one half. None was added to the other half. Rats drinking diet J_2 with C.E.E.HCl developed the full blown syndrome, but none of the rats ingesting this diet without C.E.E.HCl did.

To extend this observation, diet J_2 was prepared up to the addition of C.E.E.HCl but divided into thirds. To one third, C.E.E.HCl in the normal J_2 concentration, 0.48 percent solids was added. To another one third, C.E.E.HCl in the diet 116 concentration, 0.11 percent solids, was added. To the final third, no cysteine ethyl ester was added. At the same time, using the same batch of dietary ingredients, diet 116 was made up in a similar fashion. One preparation contained no C.E.E.HCl, another the usual 116 C.E.E.HCl 0.11 percent solids, and the third the diet J_2 C.E.E.HCl concentration, 0.48 percent solids.

As part of this same experiment, diet 116 was prepared with its usual composition but with added hydrochloric acid in the amount in diet J_2. Preparation techniques of diet J_2 were used. Data are given in Table 3-XI.

They show that cysteine ethyl ester hydrochloride is critically involved. On the other hand, it is substantially more "toxic" when fed in diet J_2 than in diet 116. This difference is not due to modes of preparation or HCl content of diets.

We then conducted experiments to determine whether cysteine hydrocholoride or cystine would induce the syndrome.

TABLE 3-XI

	C.E.E.HCl % Diet Solids	*Anemia*	*Azotemia*	*Pancreatic Lesion*
Diet J_2	0	0	0	0
	0.11	++	+	+++
	0.48	+++	++	+++
Diet 116	0	0	0	0
	0.11	0	0	+
	0.11*	0	0	0
	0.48	0	0	++

* Prepared by methods generally used to prepare diet J_2 and with added HCl to match the HCl content of diet J_2.

TABLE 3-XII

	*Body*** *wt.* (g)	*Hgb*** (g%)	*Hct*** (%)	*Pancreatic*** *Lesion*
Diet J_2				
Cysteine ethyl ester	191	9.0	30	3+
Modified Diet J_2				
Cysteine hydrochloride	240	17.3	49	0

* Sprague-Dawley rats, male, started on diets at 21–23 days of age; rats killed sixty one days later.
** Mean/ (8 animals). All differences are statistically significant, $P < 0.001$.

Data indicated that cysteine ethyl ester and cysteine hydrochloride behave in dramatically different ways. When cysteine hydrochloride was substituted in equimolar concentration for cysteine ethyl ester in liquid diet J_2, neither anemia nor pancreatic lesion developed (Table 3-XII).

Similarly, when cystine was substituted for cysteine ethyl ester, neither anemia nor pancreatic lesions developed. Solid diets were used because relative insolubility of cystine prevents its being used in liquid diets.

Cysteine ethyl ester was introduced into formulation of amino acid liquid diets by Greenstein and his colleagues because of its greater solubility than either the free amino acid or the hydrochloride. Those investigators were ". . . relying upon the powerful tissue esterases to effect the hydrolysis of these compounds to the free amino acids." (89) Little is known, however, about the metabolic fate of cysteine ethyl ester.

Adverse effects of cysteine ethyl ester may be due either to a direct effect or to an effect after conversion to another compound or to both. We thus visualize toxicity of cysteine ethyl ester as being due to either of two possibilities. Cysteine ethyl ester may react with a body constituent because of the nucleophilic properties of this mercaptan. Cysteine ethyl ester may also react with certain other dietary or endogenous constituents, as the rat's gut contents, giving rise to either antimetabolites or a non-utilizable product. An essential metabolite would then be removed from the diet. Indigenous flora may play a direct or indirect role in these processes.

In brief, these experiments show that the presence of cysteine ethyl ester but not cysteine hydrochloride or cystine is a necessary factor in the pathogenesis of the following syndrome. Hemolytic anemia, pancreatic acinar atrophy with fibrosis, and azotemia are seen. Effects of cysteine ethyl ester are conditioned by physical state of the diet, liquid being much more "toxic" than solid. Other conditioning effects are of the diet, e.g. other amino acids, and presence of the indigenous flora in the rats.

It is apparent that esters of amino acids can not be assumed to be metabolically identical to free amino acids. Further investigation of various esters and tissue esterases is in order.

Conclusion

It is clear that influences of microorganisms on mammalian metabolism and nutrition are complex and varied. They have profound effects in health and disease. Rosebury (108) has stated it well: "Our indigenous biota is, in fact, part of the environment in which we live. We need to accept its existence. Acceptance, however, need not be passive or resigned. The biota is no less subject than the rest of our environment to manipulation for human benefit."

REFERENCES

1. Pasteur, L. Cited by Gordon, H. A.: Proc. Symposium Gnotobiotic Technol., 2nd ed. University of Notre Dame Press, Notre Dame, Inc., 1959, pp. 9–18.
2. Nuttall, G. H. F., and Thierfelder, H.: Thierisches Leben Ohne Bakterien im Verdanungkanol. *Z Physiol Chem 21:* 109–21, 1895.
3. Schottelius, M.: Die Bedeutung der Darmbakterien für die Ernähiung. II. *Arch Hyg, 42:* 48–70, 1902.
4. Cohendy, M.: Expériences sur la Vie Sans Microbes. *Ann Inst Pasteur, 26:* 106–37, 1912.
5. Kuster, E.: Zentr Bakteriol. *Parasitenk, 54:* Bieheft 55–58, 1912.
6. Metchnikoff, E.: The nature of man. In Mitchell, Chalmers, P. (Ed.): *Studies in Optimistic Philosophy* (English translation). New York, Putnam, 1905.
7. Glimstedt, G.: Das Leben Ohne Bakterien, Verhandl. Anat. Ges. *Ergaenzungsheft Anat Anz., 41:* 79–89, 1932.

8. Reyniers, J. A., and Trexler, P. C.: *Micrurgical and Germfree Techniques.* Springfield, Ill., Thomas, 1943, p. 114.
9. Miyakawa, M., Iijima, S., Kishimoto, H., Kobayashi, R., Tajima, M., Isomura, N., Asano, M., and Hong, S. C.: Rearing germfree guinea pigs. *Acta Pathol Japon, 8:* 55–78, 1958.
10. Gustafsson, B. E.: Lightweight stainless steel systems for rearing germfree animals. *Ann NY Acad Sci, 78:* 17–28, 1959.
11. Trexler, P. C., and Reynolds, L. I.: Flexible film apparatus for the rearing and use of germfree animals. *Appl Microbiol, 5:* 406, 1957.
12. Levenson, S. M., and Tennant, B.: Some metabolic and nutritional studies with germfree animals. *Federation Proceedings, 22,* no. 1, Part 1: 109–119, 1963.
13. Mickelsen, O.: Nutrition—germfree animal research. *Ann Rev Biochem, 31:* 515–548, 1962.
14. Miyakawa, M. and Lucky, T. D. (Eds.): Advances in germfree research and gnotobiology. Cleveland, CRC Press. Proceedings International Symposium on Germfree Life Research, Nagoya and Inuyama, Japan, April, 1967.
15. Mirand, E. A. (Ed.): Germfree Biology. *Advances in Experimental Medicine and Biology.* New York, Plenum Press, 1969, vol. III.
16. Heneghan, J. B. (Ed.): *Proceedings of the IVth International Symposium on Germfree Research,* April 16–20, 1972, Louisiana State University, New Orleans, Louisiana. New York, Academic Press, Inc. (In Press).
17. Eijkman, C.: Eine Beri Beri-ahnliche Krankheit der Hühner. *Arch Pathol Anat Physiol, 148:* 523, 1897.
18. Cooper, E. A.: On the protective and curative properties of certain foodstuffs against polyneuritis induced in birds by a diet of polished rice. *J Hyg, 14:* 12, 1914.
19. Fridericia, L. S.: Refection, a transmissible change in the intestinal content, enabling rats to grow and thrive without B-vitamin in the food. *Skand Arch Physiol, 49:* 129–130, 1926.
20. Coates, M. E., Fuller, R., Harrison, G. F., Lev, M., and Suffolk, S. F.: A comparison of the growth of chicks in the Gustafsson germfree apparatus and in a conventional environment, with and without dietary supplements of penicillin. *Brit J Nutr, 17:* 141, 1963.
21. Osborne, T. B., and Mendel, L. B., and Feery, E.: *Feeding Experiments with Isolated Food Substances.* Washington, D.C. Carnegie Institute, Publ. 156, 1911, Pts. 1 and 2.
22. Steenbock, H., Sell, M. T., and Nelson, E. M.: Vitamin B. I. A Modified technique in the use of the rat for determinations of vitamin B. *J Biol Chem, 55:* 399, 1923.
23. Barnes, R. H., Fiala, G., and Kwong, E.: Decreased growth rate resulting from prevention of coprophagy. *Federation Proceedings, 22:* 125, 1963.

24. Briggs, G. M., Luckey, T. D., Elvehjem, C. A., and Hart, E. B.: Effect of ascorbic acid on chick growth when added to purified rations. *Proc Soc Exp Biol Med, 55:* 130, 1944.
25. Moore, P. R., Evenson, A., Luckey, T. D., McCoy, E., Elvehjem, C. A., and Hart, E. B.: Use of sulfasuxidine streptothricin and streptomycin in nutritional studies with the chick. *J Biol Chem, 165:* 437, 1946.
26. Mameesh, M. S., and Johnson, B. C.: Dietary Vitamin K requirement of the rat. *Proc Soc Exp Biol Med, 103:* 378, 1960.
27. Doft, F. S., Ashburn, L. L., and Sebrell, W. H.: Biotin deficiency and other changes in rats given sulfanilyguanidine or succinyl sulfathiazole in purified diets. *Science, 96:* 321, 1942.
28. Nielsen, E., and Elvehjem, C. A.: The growth-promoting effect of folic acid and biotin in rats fed succinylsulfathiazole. *J Biol Chem, 145:* 713, 1942.
29. Mickelsen, O.: Intestinal synthesis of vitamins in the nonruminant. *Vitam Horm, 14:* 1, 1956.
30. Stokstad, E. L. R., and Jukes, T. H.: Effect of various levels of vitamin B_{12} upon growth response produced by aureomycin in chicks. *Proc Soc Exp Biol Med, 76:* 73–76, 1951.
31. Francois, A. C., et Michel, M. C.: Mode d'action des Antibiotiques Sur la Croissance, Bibl. *Nutritio et Dieta.* New York, Karger, Basel, 1968, pp. 35–59, vol. X.
32. Dubos, R., and Schaedler, R. W.: The effect of the intestinal flora on the growth of mice, and on their susceptibility to experimental infections. *J Exp Med, III:* 407. 1960.
33. Lee, Chi-Jen, and Dubos, R: Lasting biological effects on early environmental influences. III. Metabolic responses of mice to neonatal infection with a filterable weight-depressing agent. *J Exp Med, 128:* 753–762, 1968.
34. Lee, Chi-Jen, and Dubos, R.: Lasting Biological effects of early environmental influences. IV. Notes on the physicochemical and immunologic characteristics of an entero virus that depresses the growth of mice. *J Exp Med, 130:* 955–962, 1969.
35. Gustafsson, B. E., Doft, F. S., McDaniel, E. G., Smith, J. C., and Fitzgerald, R. J.: Effects of vitamin K-active compounds and intestinal microorganisms in vitamin K-deficient germfree rats. *J Nutr, 78:* 461, 1962.
36. Wostmann, B. S. and Knight, P. Leonard: Antagonism between vitamins A and K in the germfree rat. *J Nutr, 87:* 155, 1965.
37. Levenson, S. M., Tennant, B., Geever, E., Laundy, R., Lt., and Doft, F.: Influence of microorganisms on scurvy. *Arch Intern Med, 110:* 693–702, 1962.
38. Levenson, S. M., and Rosen, H.: Unpublished data.

39. Rettger, E. A., and Johnson, L. F.: Comparative study of the nutritional requirements of Salmonella Typhosa, Salmonella Pullorum and Salmonella, *J Bact, 45:* 127, 1943.
40. Levenson, S. M., Doft, F., Lev, M., and Kan, D.: Influence of microorganisms on oxygen consumption, carbon dioxide production and colonic temperature of rats. *J Nutr, 97* (4) : 542–552, 1969.
41. Griffith, W. H., and Wade, N. J.: Choline metabolism. I. The occurrence and prevention of hemorrhagic degeneration in young rats on a low choline diet. *J Biol Chem, 131:* 567, 1939.
42. Christenson, K.: Renal changes in the albino rat on low choline and choline-deficient diets. *Arch Pathol, 34:* 633, 1942.
43. Levenson, S. M., Nagler, A. L., Geever, E. F., and Seifter, E.: Acute choline deficiency in germfree, conventionalized and open-animal-room rats: Effects of neomycin, chlortetracycline, vitamin B_{12} and coprophagy prevention. *J Nutr, 95,* (2) : 247–270, 1968.
44. Levenson, S. M., Brown, N., and Horowitz, R. E.: Dietary cirrhosis of the liver in the germfree rat. Abstracted *5th Int Congr Nutr,* Washington, D.C., 1960, p. 14.
45. De La Huerga, J., Gyorgy, P., Waldsterin, S., Katz, R., and Popper, H.: The effects of antimicrobial agents upon choline degradation in the intestinal tract. *J Clin Invest, 32:* 1117, 1953.
46. Prentiss, P. G., Rosen, H., Brown, N., Horowitz, R. E., Malm, O.E., and Levenson, S. M.: The metabolism of choline by the germfree rats. *Arch Biochem Biophys, 94:* 424–429, 1961.
47. Arnstein, H. R. V., and Newberger, A.: The effect of cobalamin on the quantitative utilization of serine, glycine and formate for the synthesis of choline and methyl groups of methionine. *Biochem J, 55:* 259, 1953.
48. Lust, G., and Daniel, L. J.: Vitamin B_{12} and the synthesis of the methyl groups of choline in *Ochromonas Malhamensis. Arch Biochem, 108:* 414, 1964.
49. Mulford, D. J., and Griffith, W. H.: Choline metabolism. VIII. Relation of cystine and of methionine to requirement of choline in young rats. *J Nutr, 23:* 91, 1942.
50. Kwong, E., Fiala, G., Barnes, R. H., Kan, D., and Levenson, S. M.: Choline biosynthesis in germfree rats. *J Nutr 96* (1) : 10–14, 1968.
51. Baez, S., and Gordon, H.: Microvascular refractoriness to epinephrine in germfree rats, symposium on gnotobiotic research, June 11–13, 1967, Madison, Wisconsin, June 11–13, 1967.
52. Nagler, A. L., Dettbarn, W-D., Seifter, E., and Levenson, S. M.: Tissue levels of acetylcholine and acetylcholinesterase in weanling rats subjected to acute choline deficiency. *J Nutr, 94* (1) : 13–19, 1968.
53. Nagler, A. L., Dettbarn, W-D., and Levenson, S. M.: Tissue levels of acetylcholine and acetylcholinesterase in weanling germfree rats subjected to acute choline deficiency. *J Nutr, 95* (4) : 603–606, 1968.

54. Nagler, A. L.: Baez, S., and Levenson, S. M.: Status of the microcirculation during acute choline deficiency. *J Nutr, 97* (2) : 232–236, 1969.
55. Rettura, G., Seifter, E., and Levenson, S. M.: The lipotropic properties of pyridyl choline 1 (2' Hydroxyethyl) pyridinium bromide. *Fed Proc, 29,* no. 2, 1970.
56. Gyorgy, P.: Antibiotics and liver injury. *Ann N Y Acad Sci, 57:* 925, 1954.
57. Rutenburg, A. M., Sonnenblick, E., Koven, I., Aprahamian, A., Reiner, L., and Fine, J.: Role of intestinal bacteria in the development of dietary cirrhosis in rats. *J Exp Med, 106:* 1, 1957.
58. Broitman, S. A., Gottleib, S., and Zamcheck, N.: Influence of neomycin and ingested endotoxin in the pathogenesis of choline deficiency cirrhosis in the adult rat. *J Exp Med, 119:* 633, 1964.
59. Baxter, J. H., and Campbell, H.: Effect of aureomycin on renal lesions, liver lipids and tissue choline in choline deficiency. *Proc Soc Exp Biol Med, 80:* 415, 1952.
60. Patek, A. P., Jr., deFritsch, N. M., and Hirsch, R. L.: Strain differences in susceptibility of the rat to dietary cirrhosis. *Proc Soc Exp Biol Med, 121:* 569, 1966.
61. Abrams, G. A., Bauer, H., and Sprinz, H.: Influence of the normal flora on mucosal morphology and cellular renewals in the ileum. A comparison of germfree and conventional mice. *Lab Invest, 12:* 355, 1963.
62. Desplaces, A., Zagury, D., and Sacquet, E.: Etude de la Fonction Thyroidienne du Rat Prive de Bacteries. *Compt Rend Hebd Seances Acad Sci (Paris), 257:* 756, 1963.
63. Windmueller, H. G., McDaniel, E. G., and Spaeth, A.: Orotic acid-induced fatty liver. Metabolic studies in Conventional and germfree rats. *Arch Biochem, 109:* 13, 1965.
64. Wostmann, B. S., Bruckner-Kardoss, E., and Knight, P. L., Jr.: Cecal enlargement, cardiac output, and 0_2 consumption in germfree rats. *Proc soc Exp Biol Med, 128:* 137, 1968.
65. Allen, D. J., and Marley, E.: Effect of sympathomimetic and allied amines on temperature and oxygen consumption in chickens. *Br J Pharmacol Chemother, 31:* 290, 1967.
66. Gordon, H. A.: Demonstration of a bioactive substance in caecal contents of germfree animals. *Nature, 205:* 571, 1965.
67. Tennant, B., Malm, O. J., Horowitz, R. E., and Levenson, S. M.: Response of germfree, conventional, conventionalized and *E. coli* monocontaminated mice to starvation. *J Nutr, 94* (2) : 151–160, 1968.
68. Levenson, S. M.: Some effects of *E. coli* and other microbes on mammalian metabolism and nutrition. *Ann N Y Acad Sci, 176:* 273–283, 1971.

69. Einheber, A., and Carter, D.: The role of the microbial flora in uremia. I. Survival times of germfree, limited flora, and conventionalized rats after bilateral nephrectomy and fasting. *J Exp Med, 123:* 239, 1966.
70. Francois, A. C.: Mode of action of antibiotics on growth. *World Rev Nutr Diet, 3:* 1–64, 1962.
71. Coates, M. E.: The mode of action of antibiotics in animal nutrition. *Chem Ind,* 1333–1335, 1953.
72. Eyssen, H., and de Somer, P.: Studies on gnotobiotic chicks: Effects of controlled intestinal flora on growth and nutrient absorption. *Ernährrungsforschung, 10:* 264–273, 1965.
73. Lev, M., and Forbes, M.: Growth response to dietary penicillin of germfree chicks and of chicks with a defined intestinal flora. *Br J Nutr, 13:* 78–84, 1953.
74. Coates, M. E., Fuller, R., Harrison, G. F., Lev, M., and Suffolk, S. F.: A comparison of the growth of chicks in the Gustafsson germfree apparatus and in a conventional environment, with and without dietary supplements of penicillin. *Br J Nutr, 17:* 1, 141–150, 1963.
75. Michel, M. C.: Metabolisme de la Flore Intestinale de Porc Degradation des Formes L et D des Acides Amines. *Ann Biol Anim Biochem Biophys, 6:* 33–46, 1966.
76. Levenson, S. M., Crowley, L. V., Horowitz, R. E., and Malm, O. J.: The metabolism of carbon-labeled urea in the germfree rat. *J Biol Chem, 234:* 2061–2062, 1959.
77. Dintzis, R. Z., and Hastings, A. B.: The effects of antibiotics on urea breakdown in mice. *Proc Natl Acad Sci, U S, 39:* 571, 1953.
78. Kornberg, H. L., Davies, R. E., and Wood, D. R.: The breakdown of urea in cats not secreting gastric juice. *Biochem J, 56:* 355, 1954.
79. Visek, W. J.: Urease immunity in liver diseases: A new approach. *Gastroenterology, 46:* 3, 326–329, 1964.
80. Sherlock, S.: *Diseases of the Liver and Biliary System,* 4th ed. Philadelphia, F. A. Davis, 1968.
81. Schwartz, S. E.: Surgical Diseases of the Liver. New York, Blakiston Division, McGraw-Hill, 1964.
82. Child, C. G., Coon, W. W., *et al.*: *The Liver and Portal Hypertension.* Philadelphia, Saunders, 1964.
83. Davidson, C. S.: *Liver Pathophysiology. Its Relevance to Human Disease.* Boston, Little, Brown and Company, 1970.
84. McDermott, W. V., Jr., and Adams, R. D.: Episodic stupor associated with an eck fistula in the human with particular reference to the metabolism of ammonia. *J Clin Invest, 33:* 1–9, 1954.
85. Doft, F., and McDaniel, E. G.: Personal Communication.
86. Seifter, E., Rettura, G., and Levenson, S. M.: Unpublished Data.
87. Seifter, E., Rettura, G., Reissman, D., Kambosos, D., and Levenson,

S. M.: Nutritional response to feeding L-phenyllactic, shikimic and D-quinic acids in weanling rats *J Nutr, 101:* 747–754, 1971.

88. Harper, A. E., Benevenga, N. J., and Wohlhueter, R. M.: Effects of ingestion of disproportionate amounts of amino acids. *Physiol Rev, 50:* 428–558, 1970.
89. Greenstein, J. P., Birnbaum, S. M., Winitz, M., and Otey, M. C.: Quantitative nutritional studies with water-soluble, chemically defined diets. I. Growth, reproduction and lactation in rats. *Arch Biochem Biophys, 72:* 396, 1957.
90. Birnbaum, S. M., Greenstein, J. P., and Winitz, M.; Quantitative nutritional studies with water soluble, chemically defined diets. II. Nitrogen balance and metabolism. *Arch Biochem Biophys, 72:* 417, 1957.
91. Birnbaum, S. M., Winitz, M., and Greenstein, J. P.: Quantitative nutritional studies with water-soluble, chemically defined diets. III. Individual amino acids as sources of "non-essential" nitrogen. *Arch Biochem Biophys, 72:* 428, 1957.
92. Winitz, M., Birnbaum, S. M., and Greenstein, J. P.: Quantitative nutritional studies with water-soluble, chemically defined diets, IV. Influence of various carbohydrates on growth, with special reference to D-glucosamine. *Arch Biochem Biophys, 72:* 437, 1957.
93. Winitz, M., Greenstein, J. P., and Birnbaum, S. M.: Quantitative nutritional studies with water-soluble, chemically defined diets. V. Role of isomeric arginines in growth. *Arch Biochem Biophys, 72:* 448, 1957.
94. Birnbaum, S. M., Greenstein, M. E., Winitz, M., and Greenstein, J. P.: Quantitative nutritional studies with water-soluble, chemically defined diets. VI. Growth studies on mice. *Arch Biochem Biophys, 78:* 245, 1958.
95. Sugimura, T., Birnbaum, S. M., Winitz, M., and Greenstein, J. P.: Quantitative nutritional studies with water-soluble, chemically defined diets. VII. Nitrogen balance in normal and tumor-bearing rats following forced feeding. *Arch Biochem Biophys, 81:* 439, 1959.
96. Sugimura, T., Birnbaum, S. M., Winitz, M., and Greenstein, J. P.: Quantitative nutritional studies with water-soluble, chemically defined diets. VIII. The forced feeding of diets each lacking in one essential amino acid. *Arch Biochem Biophys, 81:* 448, 1959.
97. Sugimura, T., Birnbaum, S. M., Winitz, M., and Greenstein, J. P.: Quantitative studies with water-soluble, chemically defined diets. IX. Further studies on D-glucosamine-containing diets. *Arch Biochem Biophys, 83:* 521, 1959.
98. Greenstein, J. P., Otey, M. C., Birnbaum, S. M., and Winitz, M.: Quantitative nutritional studies with water-soluble, chemically defined diets, X. Formulation of a nutritionally complete liquid diet. *J Nat Can Inst, 24:* 211, 1960.

99. Winitz, M., Graff, J., and Seedman, D. A.: Effect of dietary carbohydrate on serum cholesterol levels. *Arch Biochem Biophys, 108:* 576, 1964.
100. Winitz, M., Birnbaum, S. M., Sugimura, T., and Otey, M. C.: Quantitative nutritional and *in vivo* metabolic studies with water-soluble chemically defined diets. In Edsall, J. T. (Ed.): *Amino Acids, Proteins and Cancer Biochemistry,* Jesse P. Greenstein Memorial Symposium. New York, Academic Press, 1960, pp. 9–29.
101. Geever, E. F., Seifter, E., and Levenson, S. M.: Pancreatic pathology, chemically defined liquid diets and bacterial flora in the rat. *Br J Exp Pathol, 51:* 341, 1970.
102. Levenson, S. M., Kan, D., Gruber, C., Crowley, L., Jaffe, E. R., Nakao, K., Geever, E. F., and Seifter, E.: Hemolytic anemia and pancreatic acinar atrophy and fibrosis conditioned by "Elemental" liquid diets and the ordinary intestinal microflora. *Ann Surg, 174:* 469–510, 1971.
103. Levenson, S. M., Kan, D., Gruber, C., Crowley, L., Jaffe, E., Nakao, K., Geever, E., and Seifter, E.: Strange hemolytic anemia and pancreatic acinar atrophy and fibrosis. *Fed Proc, 30:* 1785–1802, 1971.
104. Levenson, S. M., Kan, D., Gruber, C., Jaffe, E., Nakao, K., and Seifter, E.: Role of cysteine ethyl ester and indigenous microflora in the pathogenesis of an experimental hemolytic anemia, azotemia, and pancreatic acinar atrophy. In Heneghan, J. B. (Ed.): *Proceedings of the IVth International Symposium on Germfree Research, April 16–20, 1972, Louisiana State University, New Orleans, Louisiana.* New York, Academic Press, Inc., 1973, pp. 297–304.
105. Reddy, B. S., Pleasants, J. R., and Wostmann, B. S.: Pancreatic enzymes in germfree and conventional rats fed chemically defined, water-soluble diet free from natural substrates. *J Nutr, 97:* 327, 1969.
106. Seifter, E., Shapiro, R., Geever, E., Nagler, A., Rosenthal, M., and Levenson, S. M.: In Miyakowa, M., and Luckey, T. D. (Eds.): *Advances in Germfree Research and Gnotobiology.* Cleveland, The Chemical Rubber Company, 1968, p. 96.
107. Shapiro, R., Rosenthal, N. A., and Gold, B. K.: Studies on the cause of a hemorrhagic syndrome in rats fed a water-soluble chemically defined diet. *J Nutr, 97:* 389, 1969.
108. Rosebury, T.: *Microorganisms Indigenous to Man.* New York, Blakiston Division, McGraw-Hill, 1962.
109. Goldberger, J., and Wheeling, G. A.: Experimental pellagra. *Arch Int Med, 25:* 451–471, (May) 1920.
110. Salmon, W. D., and Newberne, P. M.: Effect of antibiotics, sulfonamides and a nitrofuran on devlopment of hepatic cirrhosis in choline deficient rats. *J Nutr, 76:* 483–491, 1962.

CHAPTER 4

PROTEIN REQUIREMENTS IN LIVER DISEASE

CHARLES S. DAVIDSON

WE HAVE HEARD TODAY, some beautiful discussions of scientific work arising from molecular biology. This was followed by important studies in experimental animals. My discussion will center around liver disease in man, which is far less precise than molecular biology or even studies in experimental animals. As a further definition, we will be talking primarily about the protein requirements and metabolism of patients with cirrhosis, particularly when the disease arises on a setting of severe alcoholism.

Ascites is a prominent complication of cirrhosis. When large volumes are removed from the body, significant amounts of protein may be lost because the fluid contains significant quantities of protein. A second important complication of chronic cirrhosis is the altered venous circulation so that much of the portal blood from the gastro-intestinal tract by-passes the liver and enters the general circulation. In this way, the nutrients and for that matter, toxins from the gastro-intestinal tract do not go to the liver first, but reach the other tissues. Thus, amino acids absorbed from the gut, but by-passing the liver are not available for albumin synthesis or other proteins made by the liver. This situation may well contribute to protein malnutrition so characteristic of chronic liver disease. As to intestinal "toxins" it is clear that ammonia and presumably other substances arising mostly from the large intestine are causes of hepatic encephalopathy including precoma and coma. A third complication of severe cirrhosis is the chronic undernutrition which occurs and will be discussed here.

SEVERE UNDERNUTRITION

Examination of a patient with severe cirrhosis shows the characteristics of the undernutrition. Generally, subcutaneous fat is greatly diminished if not completely absent. At times it is almost impossible to find subcutaneous tissue with fat filled cells. Moreover, there is a considerable diminution in muscle mass as well. Thus, the skin is loose and often has greatly reduced elasticity.

A series of studies were made concerning the relationship between nitrogen balance and nitrogen intake in patients with cirrhosis. (1) As in similarly undernourished normal individuals, increase in nitrogen intake was followed by an increase in positive nitrogen balance, that is, an increase in nitrogen retention. The nitrogen balance data did suggest, however, that it takes rather more nitrogen intake to achieve nitrogen equilibrium and positive balance than in undernourished persons without liver disease. Loss of stool nitrogen is not different in cirrhosis from that of normal individuals or similarly undernourished individuals. It is the urine nitrogen which shows the difference. A subtle defect in nitrogen retention does exist, therefore. This is substantiated by long-term balance studies in these patients. As a patient with liver disease improves, urine nitrogen progressively decreases. Nitrogen retention therefore increases even though the patient throughout is in positive nitrogen balance and on an adequate intake.

The conclusion has been drawn in the past that undernutrition of patients with cirrhosis is caused by prolonged poor food intake. Our studies lead us to believe that undernutrition is at least partly due to some defect of intermediary metabolism leading during the height of the disease to increased urinary nitrogen loss.

Also, it has been suggested that the defect in nitrogen metabolism might be due to dietary deficiency of an essential nutrient such as choline or methionine. Studies of our patients using the nitrogen balance technique have demonstrated no further increase in positivity of nitrogen balance from the addition to the diet of choline or methionine or both.

Reversal of this undernutrition in chronic cirrhosis usually must wait until liver disease improves. Addition of moderate amounts of protein to intake may, if tolerated hasten recovery.

It has been suggested that the situation might be improved by administration of an anabolic agent such as testosterone or its congeners. When anabolic agents such as testosterone are added to the regimen in the treatment of these patients, anabolism is increased, as determined by a more positive nitrogen balance. (2) This is at the expense of urine nitrogen, not from increased food (protein) intake. Even though this does occur it is not of great clinical significance. Addition of modest amounts of extra protein to the diet can achieve the same increase of positivity of nitrogen balance, a more practical and appropriate way of achieving the desired effect.

HEPATIC COMA

It is now well known that the syndrome of hepatic precoma and coma, also known as hepatic encephalopathy, may be induced by protein feeding in susceptible patients with cirrhosis. This means that it may be necessary for these patients to limit the daily intake of protein. This situation is unfortunate for patients who may be already deficient in protein and for whom protein feeding may be an important therapeutic measure for healing the liver disease. On the other hand, encephalopathy can be reversed by administration of appropriate oral antibiotics. They reduce ammonia-forming bacteria in the gut. Lowering colon pH by administering the synthetic sugar, lactulose acts in a similar manner to decrease production of ammonia and other compounds presumed to be important in inducing the syndrome. (3) In this way, it is possible to feed adequate amounts of protein to most patients with liver disease. Exceptions are those with severe liver failure, a few who seem particularly susceptible to hepatic coma and a group of patients who develop chronic encephalopathy after surgical porta-caval anastomasis. (4,5,6)

AMINO ACIDS IN THE URINE

Aminoaciduria has been of interest in relation to liver disease since importance of liver in processing amino acids arising from dietary protein became known. These amino acids proceed via the portal vein to liver cells. Here they are accumulated in hepatocytes for synthesis of liver cell protein or plasma proteins of which albumin is quantitatively the most important. Alternatively, amino acids may be deaminated if present in excess, transaminated or passed on to the general circulation for tissue protein synthesis. How this is regulated is unknown.

When methods were developed to determine amino acids in urine, it became evident that some patients with liver disease had aminoaciduria. In certain circumstances, particularly in Wilson's disease, aminoaciduria was much greater than in simple cirrhosis. In the former situation, aminoaciduria is due to a defect in renal tubular reabsorption of amino acids, not to a high blood concentration. (7)

Aminoaciduria of cirrhosis is quite different. (8) It is far less severe than in Wilson's disease and is probably not the result of a renal defect. Occasionally a slight increase of blood amino acids above normal concentrations after a meal occurs. This is particularly true in some of the more severe liver diseases such as alcoholic hepatitis or massive liver necrosis. Thus, aminoaciduria of chronic cirrhosis is relatively small in amount and almost surely not limiting nutritionally. It is evident that the aminoaciduria is related to liver disease for, as with total nitrogen excretion, a decrease in total urinary amino acids is observed as liver disease improves. The other liver condition with marked aminoaciduria is massive liver necrosis. Here two factors presumably operate: (1) overproduction from necrosing liver tissue; (2) failure of diseased liver to accumulate or to deaminate them causing a severe aminoacidemia far exceeding the renal threshold. (9)

IMPROVEMENT IN MALNUTRITION WITH TIME

Improvement in the state of nutrition can always be expected *parri passu* with improvement in liver disease when the liver is in a situation to allow recovery such as chronic cirrhosis of the reformed alcoholic. This clear evidence is perhaps the most important aspect of malnutrition in chronic cirrhosis. This is true even of patients with massive ascites and severe undernutrition. If given a sodium restricted but otherwise normal diet, they will gradually lose their ascites and at the same time, gain in muscle mass and subcutaneous tissue. (10) Eventually the patient loses all ascites and returns to a situation indistinguishable from the normal nutritional state unless death occurs from some intercurrent event such as massive gastro-intestinal hemorrhage or severe infection. (11)

SUMMARY

The following points have been made:

1. Severe undernutrition is a common accompainment of cirrhosis.
2. There appears to be a subtle defect in nitrogen retention to which diminished food intake may contribute. It is easily reversed by protein feeding.
3. The aminoaciduria of chronic cirrhosis is not great in degree and almost surely not limiting from the point of view of nitrogen retention or loss of essential amino acids. The aminoaciduria decreases with improvement of the liver disease.
4. In chronic cirrhosis, particularly in the alcoholic, improvement can be expected, barring return to alcohol or an intercurrent illness. The improvement, however, is slower than expected for similarly undernourished individuals without liver disease. (12)

REFERENCES

1. Gabuzda, G. J., and Davidson, C. S.: Protein metabolism in patients with cirrhosis of the liver. *Ann NY Acad Sci, 57:* 776–785, 1954.
2. Gabuzda, G. J., Jr., Eckhardt, R. D., and Davidson, C. S.: Effect of choline and methionine, testosterone propionate, and dietary protein on nitrogen balance in patients with liver disease. *J Clin Invest, 29:* 566–576, 1950.
3. Elkington, S. G., Floch, M. H., and Conn, H. O.: Lactulose in the treatment of chronic portal-systemic encephalopathy. *NEJM, 281:* 408-412, 1969.

4. Gabuzda, G. J.: Ammonia metabolism and hepatic coma. *Gastroenterology, 53:* 806, 1967.
5. Gabuzda, G. J.: Hepatic coma: Clinical considerations, pathogenesis, and management. *Advances in Internal Medicine.* Chicago, Year Book, 1962, vol. II, pp. 11–73.
6. Brown, H., and Brown, Julie: Urea cycle enzymes after portacaval shunting. *JAMA, 192:* 382–384, 1965.
7. Cooper, A. M., Eckhardt, R. D., Faloon, W. W., and Davidson, C. S.: Investigation of the aminoaciduria in Wilson's Disease (Hepatolenticular Degeneration): Demonstration of a defect in renal function. *J Clin Invest, 29:* 265–278, 1950.
8. Eckhardt, R. D., Faloon, W. W., and Davidson, C. S.: Improvement of active liver cirrhosis in patients maintained with amino acids intravenously as the source of protein and lipotropic substances. *J Clin Invest, 28:* 603–614, 1949.
9. Ning, M., Lowenstein, L., and Davidson, C. S.: Serum amino acid concentrations in liver disease. *J Lab Clin Med, 70:* 554–562, 1967.
10. Eckhardt, R. D., Zamcheck, N., Sidman, R. L., Gabuzda, G. J., Jr., and Davidson, C. S.: Effect of protein starvation and of protein feeding on the clinical course, liver function, and liver histochemistry of three patients with active fatty alcoholic cirrhosis. *J Clin Invest, 29:* 227–237, 1950.
11. Davidson, C. S.: Cirrhosis of the liver treated with prolonged sodium restriction. Improvement in nutrition, hepatic function and portal hypertension. *JAMA, 159:* 1257–1261, 1955.
12. Davidson, C. S.: *Liver Pathophysiology. Its Relevance to Human Disease.* Boston, Little Brown & Co., 1970.

CHAPTER 5

AMINO ACID REQUIREMENTS AND PLASMA AMINO ACIDS

A. E. Harper

New information about protein and amino acid requirements accumulates slowly—more slowly than reviews of existing information. Nevertheless, this is an appropriate time to reassess knowledge of the nutritionally indispensable amino acids as two recent efforts to evaluate information on the subject and identify unresolved problems should be published soon. A second edition (1) of the bulletin "Evaluation of Protein Nutrition" (2) is being prepared by the Amino Acid Committee of the Food and Nutrition Board of the NAS-NRC.* A new bulletin on "Protein and Energy," (3) which represents a revision of the previous FAO/WHO** protein reports (4,5) is being prepared by an expert committee. The information on nitrogen and amino acid requirements presented here will be reviewed with a somewhat different perspective from those reports.

Man requires nitrogen and several specific amino acids for the synthesis of tissue proteins and as precursors of a variety of other nitrogenous constituents of tissues. In 1930 Terroine (6) emphasized the dual nature of the nitrogen requirement and distinguished between the specific requirement for indispensable (essential) amino acids and the nonspecific requirement that can be met by a variety of nitrogenous compounds, including ammonium salts. Nitrogen is

* National Academy of Science—National Research Council (U.S.).

** Food and Agriculture Organization/World Health Organization (United Nations).

lost throughout life in urine, stool, sweat, sloughed skin, hair, nails and in various other body secretions and excretions. Amino acids are not stored in the body. As a result both nitrogen and the specific indispensable amino acids must be supplied continuously by the diet, not only for growth of the infant, but also for maintenance of the mature individual. Amino acid and total nitrogen needs of the infant and of physically mature men and women have been quantified. Changes in these requirements with age have been documented and several factors that influence nitrogen and amino acid utilization have been identified. I shall review these topics briefly and also say a little about plasma amino acid concentrations and their regulation and how changes in dietary amino acid intake may be reflected by changes in the composition of tissue amino acid pools.

NITROGEN AND AMINO ACID REQUIREMENTS

Nitrogen Requirement

The nitrogen requirement of man has been estimated regularly for decades. Observations by Voit, Atwater, and others during the latter part of the last century and the early part of this, that working men consumed regularly about 125 gm of protein a day, (7) led to the assumption that adult man needed of the order of 20 gm of nitrogen daily. At the same time as these epidemiologic observations were accumulating, Siven, Chittenden, and Voit, himself, demonstrated that adult man could be maintained in good health and nitrogen equilibrium with an intake of somewhat above 4 and somewhat below 6 gm of nitrogen (equivalent to 25–40 gm of protein) daily. (7) The accumulated information from nitrogen balance studies since then has not defined the requirement much more exactly. From many reports (1,3) 70 mg of nitrogen/kg of body wt/day (about 0.45 gm of protein) represents a reasonable average value for the amount of nitrogen from diets containing good quality proteins or from mixed diets needed to maintain adult man in nitrogen equilibrium. This criterion was proposed by Terroine (6) as most valid for estimation of the requirements of adults.

Use of the nitrogen balance procedure for assessing nitrogen requirements has three obvious limitations. First, nitrogen intake tends to be overestimated and excretion to be underestimated solely as a result of methodological problems. (8) Errors are then additive leading to underestimation of the requirement. In addition, nitrogen lost from skin, hair, secretions, and other minor routes is rarely measured. A positive nitrogen balance of about 0.5 gm/day, therefore, is required to ensure that an adult subject is in nitrogen equilibrium. (9) Second, the nutritional value of a protein is a function of its amino acid composition. (10) Hence, more of a protein of low nutritional quality (i.e. with an unbalanced pattern of indispensable amino acids) * must be consumed to meet the specific amino acid requirements than protein of high nutritional quality. Bricker *et al.* (11) demonstrated that young women needed more flour protein than milk protein to maintain nitrogen equilibrium.**

---

* The indispensable (essential) amino acids are histidine, isolencine, leucine, lysine, methionine, phenylalanine, threonine, tryptophan, and valine.

The dispensable (non-essential) amino acids are alanine, arginine, aspartic acid, cystine, glutamic acid, glycine, proline, serine, and tyrosine. They can be synthesized in the body. Cystine can be synthesized only from methionine and tyrosine only from phenylalanine. As their respective precursors are spared when cystine and tyrosine are present in the diet, these two amino acids are sometimes termed semi-indispensable.

Cystine and tyrosine are not readily synthesized in sufficient amounts by premature infants, so may be indispensable for them. Arginine is not synthesized in sufficient amounts for rapid growth by the young of many species, but is by the human infant.

** Estimation of the nitrogen requirements of the subjects of Bricker *et al.* (11) has been done by interpolation. They provide values for the amount of nitrogen required to maintain nitrogen equilibrium. By extrapolation of their regression analysis, values were calculated to allow for positive balance of 1.0 gm of nitrogen/day to cover skin losses and adult growth. The concept of adult growth is questionable. (1) These two values, therefore, have been averaged to give a requirement for positive balance of about 0.5 gm of nitrogen/day to allow for all unmeasured minor nitrogen losses and positive errors of nitrogen balance procedures. The value of 76 mg of nitrogen/kg of body weight/day obtained by this procedure for milk protein corresponds quite well with the average value of 70 mg/kg/day estimated from serveral studies. (1,3)

Addition of lysine to the flour improved its nutritional value. An average nitrogen requirement is thus meaningful only when the amount of nitrogen recommended is accompanied by the required quantities of the indispensable amino acids. Third, when protein intake is decreased, subjects tend to come into nitrogen equilibrium in time with lower nitrogen intakes. The point at which protein depletion begins is, therefore, difficult to identify. In experiments in which subjects are consuming gradually decreasing amounts of protein for some time, adaptation to a low protein intake may occur during the course of the study. Nevertheless, despite these shortcomings, (8,12) there is no adequate substitute for the nitrogen balance procedure for estimating nitrogen and amino acid requirements of adult man.

Nitrogen Losses

Summation of nitrogen losses of subjects consuming no protein has been proposed as an alternative approach for estimating nitrogen requirements. This procedure, which has been used by FAO/WHO committees (4,5) involves measurement of urinary, fecal, and integumental losses of nitrogen by subjects consuming a nitrogen-free diet.

Urinary nitrogen excretion falls rapidly for a few days when protein is withdrawn from the diet, then remains relatively constant for several days. In view of the difficulty of identifying an inflection point as the curve for nitrogen excretion approaches a plateau, the plateau value is taken as the obligatory urinary nitrogen loss. An average value of 1.3 mgN/basal Kcal, or 34 mg/kg of body wt for a 70 kg man with a basal metabolism of 1850 Kcal/day has been estimated from accumulated information. (1) This corresponds well with values of 37–38 mg/kg reported for more recent studies by Young and Scrimshaw (13) and Calloway and Margen. (14) As several days of protein depletion occur before urinary nitrogen excretion approaches constancy, these values are likely to be underestimates. They are used more because they are reproducible than because they are known to represent usual endogenous losses.

Obligatory fecal nitrogen output of subjects fed a protein-free diet is quite variable. The average for values reported by various investigators is 12 mg/kg of body wt/day, (1) not appreciably different from those reported in recent studies. (13,14)

Nitrogen losses from skin and other integumental sources have been estimated at 4 mg/kg of body wt/day (1,2) and may be less. (15) Other minor losses probably amount to 2–3 mg/kg of body wt/day. (16)

Summation of these average losses, as shown in Table 5-I, gives a value of 52 mg of nitrogen/kg of body wt/day. This would represent 0.32 gm of protein/kg of body wt/day or 3.6 gm of nitrogen, equivalent to 23 gm of protein, for a 70 kg man. These are average values and should be compared with the values of 70 mg of nitrogen and 0.45 gm of protein/kg of body wt/day estimated for average requirements on the basis of nitrogen balance studies.

If the sum of nitrogen losses is taken as an estimate of the average minimum nitrogen requirement, this assumes 100 percent efficiency of utilization of dietary nitrogen. Although measurements of biological value indicate that utilization of high quality proteins may approach 100 percent when fed in limiting amounts, especially to young, growing animals, (10) response per unit of nutrient tends to fall as the requirement is approached. (17,18) This pattern is characteristic of biological processes so it would be unusual if efficiency of nitrogen utilization did not fall as the requirement was approached and if the requirement did not exceed the sum of nitrogen losses. Measurements of amounts of high

TABLE 5-I

Average Obligatory Nitrogen Losses of Adult Men Consuming a Protein-Free Diet

	Nitrogen loss (mg/kg body wt/day)
Urine	34
Stool	12
Cutaneous	4
Minor losses	2
Total	52

quality proteins needed to maintain adult man in nitrogen equilibrium, obtained in nitrogen balance experiments, commonly yield values for efficiency of protein utilization of 65-75 percent. (11,14,19)

Minimum Nitrogen Intake to Maintain Nitrogen Equilibrium

Results of a study by Rose and Wixom (20) are instructive in relation to minimum nitrogen requirements. They studied young men consuming a highly purified diet containing indispensable amino acids in proportions in which they are required. The subjects had positive nitrogen balances of about 0.15 gm/day with nitrogen intakes of only 3.5 gm daily, much of it from nonspecific sources. Let us assume that minor unmeasured nitrogen losses were about 300 mg/day. This together with an allowance for positive errors of the nitrogen-balance technique would require a positive nitrogen balance of 0.5 gm/day to ensure nitrogen equilibrium. Subjects were then in negative nitrogen balance of only 0.35 gm/day. This would represent a loss of body protein of only 66 gm in 30 days and, assuming a body protein content of 18 percent, of 365 gm of body wt. However, as shown in Table 5-II, to achieve a positive nitrogen balance of 0.5 gm/day a nitrogen intake of between 4 and 6 gm/day was required, a figure quite in line with those from other nitrogen balance studies with high quality proteins. Intakes of energy in this study were high, 55 Kcal/kg/day, and should have assured high efficiency of nitrogen utilization. Also, subjects were fed

TABLE 5-II

Estimate of Nitrogen Requirement of Men Consuming Amino Acid Diets

N-intake	*N-balance*	*Increase in N-retention*	*Efficiency of retention of increment**
gm/day	gm/day	gm/day	%
3.5	+0.15		
4.0	+0.29	0.14	28
6.0	+0.46	0.31	8.5

Data from Rose and Wixom (20)

* Calculated from $\frac{\text{increment in N-retention}}{\text{increment in N-intake}} \times 100$ e.g. $\frac{(6.0-4.0)}{(0.46-0.29)} \times 100 = 8.5$

gradually decreasing amounts of nitrogen over several weeks so may have undergone some adaptation to low protein intake. This analysis of the results from other nitrogen balance studies with diets containing protein supports the conclusion that nitrogen, even from high quality sources, is not used efficiently by the body as the requirement is approached.

Amino Acid Requirements

The two major studies of amino acid requirements of adults are those of Rose and his associates on young men (21) and those of Leverton and her associates on young women (22) As well, there have been several studies by others of requirements for one or more of the amino acids. These have been tabulated by Munro. (23) Hegsted (9) has evaluated these studies and has estimated requirements for each of the indispensable amino acids by analyzing statistically the relationship between amino acid intake and N-retention using all of the individual values for women. The estimated requirements for adults, listed in Table 5-III, are those of Rose and associates (21) for men and those calculated by Hegsted (9) for women. The values for men tend to be high, as Rose and associates accepted as the requirement the highest of the values for the individuals they studied. The values for women represent closer to average requirements.

TABLE 5-III

Amino Acid Requirements of Adult Man

	Mg/day		*Mg/kg/body wt/day*		
	Male (70 kg)[1]	*Female (58 kg)*[2]	*Male*	*Female*	*Average*
Ile	700	550	10.1	9.5	9.8
Leu	1100	725	15.7	12.6	14.1
Lys	800	545	11.4	9.3	10.3
Met		700		12.1	
+Cys	1010		14.4		13.2
Phe	1100	700[3]	15.7	12.1	13.9
Thr	500	375	7.1	6.5	6.8
Trp	250	168	3.6	2.9	3.2
Val	800	622	11.4	10.7	11.0
Total	6260	4385			82.3

[1] Rose (21)
[2] Hegsted (9)
[3] Burrill and Schuck (24)

Although within each study little relationship was observed between requirement and body weight, when values for men and women are expressed per kg of body weight, differences between sexes are small. This suggests that if body weight differences are large a relationship between requirement and body weight can be demonstrated. In view of wide variability among values for individuals (9,21–23) and different criteria used for estimation of requirements differences between sexes are of doubtful significance.

Averages of commonly accepted values for men and women given in the last column of Table 5-III are probably reasonable estimates of average adult requirements for indispensable amino acids. They are surprisingly small. Even assuming a coefficient of variability of 15 percent (3) and therefore addition of 30 percent to them to cover needs of individuals with highest requirements, they would still be surprisingly small. They would be less than the safe intake of double the estimated minimum requirement proposed by Rose, (21) but would be above the highest requirement he and his associates observed for any one individual. Criterion for nitrogen equilibrium, however, in studies of amino acid requirements has usually been attainment of a nitrogen balance of zero. It is, therefore, very likely that all of these figures are underestimates. Hegsted (9) calculated amino acid requirements for maintenance of positive balance of 0.5 gm of nitrogen per day and derived much higher values.

Amino Acid Requirements Estimated from Minimum Protein Intakes to Maintain Nitrogen Equilibrium

The average nitrogen requirement of adult man can be met with about 450 mg of protein/kg of body wt/day, and and total indispensable amino acid requirements are only about 80 mg/kg of body wt/day. If these figures are valid then the amount of nitrogen required from indispensable amino acids would be less than one-fifth of the total nitrogen requirement. Since about half the nitrogen of high quality proteins is from indispensable amino acids it should be possible to dilute such proteins with some nonspecific nitrogen source and still meet both specific amino acid and total nitrogen requirements without altering nitrogen intake.

Swendseid and associates (25,26) fed egg protein diluted with dispensable amino acids or other nonspecific nitrogen sources to human subjects. They were able to keep some subjects in nitrogen equilibrium with from 8 to 10 gm of egg protein as the source of indispensable amino acids using diets that provided at least 6.5 gm of total nitrogen daily. Scrimshaw and associates (27–29) also observed that nitrogen retention of subjects consuming high quality proteins in about the amount needed to meet their nitrogen requirement was not impaired when the protein was diluted 20 to 30 percent with nonspecific nitrogen sources. The dilution tolerated was less than that observed by Swendseid and associates (25, 26) but total nitrogen provided was marginal, about 4.5 gm/day. These studies demonstrate that total nitrogen becomes limiting before indispensable amino acids for adult man when amounts of high quality protein consumed meet nitrogen requirements. Snyderman *et al.* (30) have shown that nitrogen retention of infants is not impaired when milk protein is diluted with nonspecific nitrogen compounds such as glycine and urea and have suggested that as milk protein intake is decreased total nitrogen becomes limiting before specific amino acids even for the infant. The degree of dilution at the requirement level is not readily estimated from their experiments but would appear to be close to 20 percent.

Calculation of quantities of indispensable amino acids contained in diets of Swendseid *et al.* (25) provides a method of estimating amino acid requirements of adults. Calculated values are shown in Table 5-IV together with estimates of requirements determined using amino acid diets (Table 5-III). Agreement between the two sets of values, expressed per kg of body wt, is surprisingly good except for the sulfur-containing amino acids and tryptophan. This suggests the need for reinvestigation of requirements for these two amino acids. Fisher and associates (31) have recently reported lower values for tryptophan requirements than have been accepted previously and requirements of most subjects studied

TABLE 5-IV
Adult Amino Acid Requirements and Minimum Amounts of Indispensable Amino Acids from Egg Protein that will Maintain Nitrogen Equilibrium in Some Individuals[1]

	Male		*Female*	
	Requirement	*10.0 gm of egg protein*[2]	*Requirement*	*8.1 gm of egg protein*[2]
	mg/day	mg	mg/day	mg
Ile	700	623	550	511
Leu	1100	881	725	716
Lys	800	698	545	569
Met + cys	1010	579	700	471
Phe + tyr	1100	989	700	803
Thr	500	512	375	416
Trp	250	149	168	121
Val	800	685	622	556
Total	6260	5116	4385	4163

[1]Swendseid et al.[25,26]
[2] Diets provided at least 6.5 gm of nitrogen daily, largely from nonprotein sources.

by Rose and associates did not exceed 150 mg/day. Hegsted (9) has discussed problems in estimating requirements for sulfur-containing amino acids and has suggested that requirements for these are probably overestimated.

It should be emphasized that some subjects studied by Swendseid *et al.* (25) were not in nitrogen equilibrium with these low amounts of egg protein. Minimum amino acid intakes from the amounts of several proteins needed to maintain adult subjects in nitrogen equilibrium (1) are usually considerably higher than figures listed in Table 5-IV. Also, estimates (9) of amounts of amino acids required to maintain adult subjects in positive nitrogen balance of 0.5 gm/day given in Table 5-V (column 4) are two, three, or more times average requirements from Table 5-III and for most amino acids exceed considerably upper limits of the 95 percent confidence range of requirements for women and the values of Rose for men. Unfortunately, accuracy of these estimates (column 4) is limited as far fewer subjects have been studied in upper than lower ranges of amino acid intakes. This table nevertheless emphasizes that commonly accepted amino acid requirements (Table 5-III, column 5) are probably underestimates. Values in the last column of Table 5-V represent upper limits of the 95 percent confidence range.

TABLE 5-V

Estimated Amino Acid Requirements of Adults

	Average from Table 5-III	*Hegsted, upper 95% confidence limit*[1]	*Rose*[2]	*Hegsted, to ensure +0.5 gm N-retention*[1]	*Average column 1 +30% and column 2*
	(all values reported as mg of amino acid/kg body wt/day)				
Ile	9.8 (13.1)[3]	14.0[4]	10.1	28.3	13.6
Leu	14.1 (18.3)	15.7	15.7	43.3[6]	17.0
Lys	10.3 (13.4)	13.9	11.4	29.2	13.6
Met + cys	13.2 (17.1)	12.2[5]	14.4	14.6[5]	14.6
Phe + tyr	13.9 (18.1)	20.2[5]	15.7	24.2[5]	19.2
Thr	6.8 (8.4)	8.6	7.1	29.2[6]	8.5
Trp	3.2 (4.2)	3.2	3.6	4.5	3.8
Val	11.0 (14.3)	11.8	11.4	16.6	13.0
Total	82.3 (106.9)	99.6	89.4	189.9	103.3

[1] Assuming 60 kg body wt.
[2] Assuming 70 kg body wt.
[3] Figures in parentheses represent average +30% to allow for individual variability.
[4] Upper limit for isoleucine indicated that the distribution of requirements was badly skewed. The value has therefore been adjusted downward to keep it more in line with the distribution pattern for the other amino acids.
[5] 500 mg of total sulfur amino acids from cystine.
900 mg of total aromatic amino acids from tyrosine.
[6] The slopes of the regression lines for these amino acids were low.

Requirements of Infants and Adults

Amino acid and nitrogen requirements are high at birth and fall sharply as the phase of rapid growth is passed. Hegsted (32) has calculated change with age in protein requirements of man from estimates of requirements for growth and maintenance. Calculations show that after the first year of life requirement for growth is much smaller than that for maintenance. Hartsook and Mitchell (33) have made similar estimates of the pattern of change with age for the total nitrogen and total sulfar-containing amino acid requirements of the rat. Their results indicate that amino acid requirements fall more sharply than the total nitrogen requirement.

The nitrogen requirement of infants is estimated from amounts of milk or formula required to maintain a satisfactory growth rate. For breast-fed infants, 385 mg of nitrogen/kg of body wt/day (equivalent to 2.4 gm of protein) is a commonly accepted value. (3) Foman *et al.* (34) have recently reported values slightly below this, 345 mg/kg/day (2.15 gm of protein/kg/day) for male infants fed a well-balanced formula. Amino acid requirements of infants can be estimated by calculating amounts of amino acids provided by quantities of milk or formula diets that support satisfactory growth rates. (1) Also, infant requirements for most of the amino acids have been determined directly by Holt, Snyderman, and associates (35) using highly purified diets containing crystalline amino acids.

A comparison of the amino acid requirements of infants and adults is given in Table 5-VI. The lowest value for each amino acid from work of Holt *et al.* (35) or from calculations based on the amino acid content of milk or formula diets in amounts that supported satisfactory growth (1) has been used in compiling values tabulated for the infant.

Adult values are those from the last column of Table 5-V. The column giving the infant/adult ratios indicates that adult requirements for most amino acids are one-sixth to one-eighth of those for the infant. Requirements for isoleucine, sulfur-containing amino acids, and tryptophan fall less with age than others. The last two are amino acids for which adult requirements determined using amino acid diets are high compared to the minimal intakes from egg protein listed in Table 5-V. The isoleucine requirement is probably overestimated despite adjustment made in the upper 95 percent confidence limit of Hegsted. (9) Low ratios for the other two suggest either that average adult requirements for

TABLE 5-VI

Amino Acid Requirements of Infants and Adults

	Infant[1]	*Adult*[2]	*Infant/Adult*
	mg/kg body wt/day		
His	33	?	?
Ile	83	13.6	6.1
Leu	135	17.0	7.9
Lys	99	13.6	7.3
Met + Cys	58	14.6	4.0
Phe + Tyr	141	19.2	7.3
Thr	68	8.5	8.0
Trp	21	3.8	6.1
Val	92	13.0	7.1
Total	730	103.3	

[1] Lowest value from milk or formula diets or from amino acid mixtures that support a satisfactory rate of gain.[1,3]

[2] Estimates from Table 5-V, column 5.

tryptophan and sulfur-containing amino acids are high, as concluded by Fisher (31) and Hegsted, (9) or that requirements for other amino acids are low as would be concluded from information in Column 4, Table 5-V. More likely, from columns 1 and 4 Table 5-V, is that accepted values for these three amino acids are closer to maximum than average values.

An important difference between amino acid needs of the infant and adult is emphasized by Table 5-VI. Histidine is listed as indispensable for the infant (35) but not for the adult. (21) Although Rose and Wixom (20) were able to maintain adult men in nitrogen equilibrium for 60 days on a diet devoid of histidine, there is no evidence that it is synthesized by mammals. It is obviously well conserved by adult man but should, nevertheless, be considered an essential nutrient unless evidence that the adult can synthesize it is obtained. (1)

Arginine, which is required by most species because it cannot be synthesized rapidly enough, is not a dietary essential for growth of the human infant. (35)

Information on average total nitrogen and indispensable amino acid nitrogen requirements of infants and adults, Table 5-VII, permits further examination of changes that occur with maturation. Infant/adult ratios reveal that requirements for indispensable amino acids fall more sharply than the requirement for total nitrogen. This means, as illustrated

TABLE 5-VII

Amino Acid and Total Nitrogen Requirements of Adults and Infants

	Nitrogen *mg/kg/day*	*Indispensable Amino Acid N* *mg/kg/day*	$N_i/N_t \times 100$[2]
Adult	90[1]	20[1]	22
Infants	385	152[1]	39
Infant/Adult	4.3	7.6	

[1] These values represent the average value + 30% to cover the upper range of individual variability, making them comparable to the nitrogen requirement of the infant which represents the upper limit of the requirement. Amino acid requirements for infants from Table 5-VI, column 1 and for adults from Table 5-V, column 5.

[2] N_i = nitrogen needed from indispensable amino acids (Column 2).
N_t = total nitrogen requirement (Column 1).

by indispensable nitrogen to total nitrogen ratios in column 3, that the adult requires a smaller proportion of indispensable to total amino acids than the infant. The proportion of 40 percent of indispensable amino acid nitrogen for the infant is less than the proportion provided by high quality proteins such as milk and egg. This could account for observations of Snyderman *et al.* (30) that some dilution of milk proteins with nonspecific nitrogen, such as glycine and urea, does not impair the nutritional value of milk for infants. Even low quality proteins, such as those of cereal grains, contain more than 22 percent of indispensable amino acids. The amino acid pattern of such proteins, however, is frequently unbalanced.

Amino Acid Requirements and Protein Quality

It is difficult to account for differences in nutritional value of proteins for adults if amino acid requirements are as low as average values of Table 5-III, even when the amino acid pattern of the diet is quite unbalanced. In the study of Bricker *et al.*, (11) however, women weighing 59 kg required about 7.76 gm of nitrogen from flour to maintain nitrogen equilibrium. This amount of flour protein would provide 870 mg of lysine, (36) well above the calculated average requirement of 720 mg/day from Table 5-III. On the other hand, it is well below the value calculated by Hegsted (9) to ensure positive nitrogen balance of 0.5 gm/day. It is also just 50 mg above the value listed in column 5 of Table 5-V, proposed as a more realistic requirement. These calculations suggest that when flour is the protein source, excess nitrogen must be consumed to meet the requirement for lysine, the limiting amino acid. The amount of total sulfur-containing amino acids provided with the 4.5 gm of nitrogen from milk required by these subjects would be 935 mg, about 60 mg above the calculated requirement (Table 5-V). Thus, when milk is the protein source nitrogen and limiting amino acids, those containing sulfur, are provided in about the right proportions but lysine (2200 mg) and most other amino acids are in excess.

SOME FACTORS AFFECTING NITROGEN AND AMINO ACID REQUIREMENTS

Many factors affect nitrogen and amino acid utilization. Reduced efficiency of utilization of nitrogen and amino acids should not necessarily be equated with increased requirements. Amino acids may be diverted into catabolic pathways as the result of an energy deficit or a metabolic shift owing to a pathologic state. This diversion may be counterbalanced, or partly so, by increased protein or amino acid intake even though the primary problem is not an increased requirement for nitrogen or amino acids.

Work, High Temperature and Sweating

Evidence that protein requirement is increased by work *per se* is not convincing. Muscle mass increases during adaptation to constant physical activity, but the amount of extra protein needed for this cannot be large. Work, however, increases caloric requirements. If energy intake is not increased sufficiently, amino acids may be diverted into energy-yielding rather than synthetic reactions.

Nitrogen is lost in sweat. Either heavy work or a hot environment, both of which increase sweating, can increase this loss. Heavy work in a hot environment could increase it substantially (Holmes, E. G., p. 255). The amount of nitrogen lost in sweat by adult males has been variously estimated from 150–360 mg/day. Profuse sweating may increase the rate of loss several-fold, at least over short periods without compensation by decreased urinary loss. Continued exposure to such conditions, however, does result in adaptation decreasing the amount of nitrogen lost in sweat. (3,37) Whether nitrogen lost in sweat represents specific amino acids is not known.

Amino Acid Proportions—Relation to Adaptability

Both Romo and Linkswiler (38) and Clark *et al.* (39,40) have observed that nitrogen balance by adult subjects consuming about 6.0 gm of nitrogen daily becomes increasingly positive as the proportion of nitrogen from indispensable amino acids is increased. In these studies, with a well-balanced dietary pattern of indispensable amino acids 30 percent (40) or 40 percent (38) of nitrogen was required from indispensable amino acids to attain nitrogen equilibrium. Swendseid, (25,26) in contrast, found that only 6 to 13 percent of nitrogen in diets containing 6.5 or 13 gm of nitrogen was required from indispensable amino acids of egg. Here, too, nitrogen retention increased as the amount of indispensable amino acids was increased. In experiments of Romo and Linkswiler (38) nonspecific nitrogen was decreased as the proportion of amino acids from indispensable amino acids was increased. Scrimshaw and associates (27–29) were able to dilute high quality proteins 25 to 30 percent with nonspecific nitrogen without impairing nitrogen retention when nitrogen intakes were only 4.5 gm/day. In a different type of study, Kies *et al.* (41) and Snyderman *et al.* (30) have reported that nitrogen retention by subjects consuming marginally adequate amounts of protein is improved if they are given supplements of nonspecific nitrogenous compounds.

Few studies of this type have been described. Results suggest that amino acid requirements are influenced by total nitrogen, the pattern of indispensable amino acids and the proportion of indispensable to dispensable amino acids in the diet. It is tempting to speculate that with low to moderate nitrogen intakes (4.5 to 6.0 gm) requirements for indispensable amino acids are greater than when total nitrogen intake is high. (40,42) Also, when the pattern of amino acids in unbalanced, the requirement for the limiting amino acid is elevated. Studies on animals cited by Munaver and Harper (43) indicate in fact that the requirement for the limiting amino acid may be elevated when the dietary amino acid pattern is out of balance.

Several amino acids tend to be conserved by the animal body when they are not provided in the diet. (44) Also, man and animals, after a period of adaptation, achieve nitrogen equilibrium with low protein intakes. (45) In protein deficient subjects not only may there be conservation of specific amino acids (44) but also reduced rates of tissue, especially muscle, protein turnover. (46) The adaptive processes are not well understood, and the extent to which they come into play before a state of depletion is reached is not clear. If such reactions are initiated when intake of indispensable amino acids is low, but not deficient, the degree to which a dietary protein can be diluted with nonspecific sources of nitrogen would depend upon duration of the experiment and adaptive capacity of the individual. These factors might be highly variable. (47) Investigations of influence of adaptability on amino acid and nitrogen requirements and of the point at which adaptation shades into deficiency are needed to provide a better understanding of the relationship between amino acid composition of diet and requirements.

Energy Intake

Energy intake is perhaps the most important factor that influences nitrogen and amino acid utilization in healthy persons. The subject has been reviewed by Munro. (48) Carbohydrate has a unique sparing action on body protein during starvation, probably because fat cannot serve as a precursor of the glucose needed by the body. Inadequacy of total energy intake, nevertheless, leads to utilization of amino acids for energy with the result that nitrogen retention decreases. Even high nitrogen intakes will not restore nitrogen equilibrium if the energy deficit is too great. (49) This effect can be demonstrated for lysine specifically (50) and is probably true for other amino acids. Although an energy deficit results in amino acid wastage, it is probably inaccurate to equate the reduced efficiency of amino acid utilization from this with elevated amino acid requirements.

Rose and associates (51) observed that subjects consuming amino acid diets needed more energy to maintain nitrogen equilibrium than subjects consuming equivalent amounts of nitrogen from intact proteins. Much nitrogen in the amino acid diets they used was from glycine and urea. More recently Anderson and Linkswiler (52) have found that this does not appear to be true if all of nitrogen is from a balanced mixture of amino acids. Their work does suggest, however, that when a large proportion of nitrogen is from such compounds as glycine and diammonium citrate, energy needs may be increased. Swendseid *et al.* (26) have also observed that a mixture of the dispensable amino acids is used more efficiently than glycine and diammonium citrate.

Trauma, Stress, and Infection

Trauma, severe stress, and infection increase urinary nitrogen losses. (53,54) Surgical trauma induces a protein catabolic state equal to that of starvation, and unanesthetized trauma and sepsis following surgical trauma can induce an even greater degree of protein catabolism.

Severity and duration of nitrogen loss vary with degree of insult. Reaction in these conditions is highly variable, but the catabolic state can be counterbalanced in large measure in some by intravenous hyperalimentation. (53) It is debatable whether nitrogen losses in these conditions should be equated with increased requirements for nitrogen and amino acids. It seems more appropriate to consider them as pathologic rather than as nutritional problems even though they may respond to nutritional treatments. Nevertheless, nitrogen and amino acid requirements are increased during the period when depleted tissues, regardless of the cause of depletion, are being repleted. (55)

PLASMA AMINO ACID CONCENTRATIONS AND REGULATION OF AMINO ACID METABOLISM— A MAMMALIAN MODEL

Amino Acid Pools

Amounts of amino acids in organ and tissue pools, as shown for the young, growing rat (Table 5-VIII), are small in relation to amino acid requirements. (56) Muscle provides the largest reserve of free amino acids; for threonine this represents 18 percent of the daily requirement but for leucine and valine only 1.0–1.5 percent of the daily requirements. There is no relationship between the amounts of amino acids in the free pools and the amounts needed by the body. The plasma

TABLE 5-VIII

Comparison of Amino Acid Pools and Requirements of Rat
(μmoles/100 gm body wt)

	Plasma	*Intestine*	*Liver*	*Muscle*	*Requirement*
Threonine	1.1	2.8	1.6	72	400
Lysine	1.1	3.9	1.7	32	600
Histidine	0.3	1.2	3.1	33	140
Valine	0.8	3.4	1.5	8	500
Leucine	0.5	5.7	2.6	6	500

pool is small in relation to the total body pool. It contains about 0.25 percent of the amount of threonine required and about 0.1 percent of the amount of leucine required daily. Ingestion of an amount of an amino acid mixture that just met the requirements of the rat would represent an influx of from 400 to 1000 times the amounts in the plasma pool. For the mature animal the figures would be smaller as requirements would be less and body pools would be larger.

In adult man the plasma pool is estimated to contain about one gram of amino acids (57) with indispensable amino acids accounting for 300 mg. (58) Although the total amount of indispensable amino acids required to meet the average daily requirements of adult man is only about 6.0–7.0 gm (Table 5-V), average intakes in rich countries are commonly 6 or more times this. Assuming the plasma pool remained constant, this would represent 120 to 140 times the amount of amino acids contained in it. Thus the plasma pool would turnover every 10 to 12 minutes on the average, but the rates for individual amino acids would differ considerably.

Evidence that amino acid concentrations in plasma are regulated accurately comes from observations that fasting concentrations fall only slowly and not very greatly even during a prolonged fast. (59) Further, in healthy, well-nourished subjects or animals plasma amino acid concentrations return rapidly toward a standard value after they have been elevated owing to ingestion or injection of an amino acid load. (60) Also, plasma amino acid concentrations do not vary greatly over time in healthy individuals consuming a normal diet. This implies accuarate homeostatic or feedback regulation with mechanisms for monitoring body fluid amino acid concentrations. The mechanism would detect deviations from a standard value. It would also do work necessary, after detection of a deviation, to remove a surplus or compensate for a deficit.

Regulation of Amino Acid Metabolism

Biological systems have relatively few basic regulatory mechanisms. Changes in enzyme activity and changes in the rate of transport of nutrients and metabolites represent basic mechanisms common to all organisms. Without change in quantity of enzyme protein, rate of an enzymatic reaction is influenced by changes in concentration of substrate, cofactors, activators, or inhibitors. As well, many enzymes in higher organisms are adaptive, *i.e.* amount of enzyme may be altered by nutritional or endocrine changes influencing rates of enzyme synthesis or degradation. With transport which, for many nutrients, means active, carrier-mediated transport the response is similar. Without change in quantity of carrier, rate of transport is influenced by changes in concentrations of solute, cofactors, activators or inhibitors. Other systems also integrate and coordinate metabolism in various tissues and organs of higher organisms. These systems, the endocrine system and the central nervous system, nevertheless, still exert their effects through regulation of the basic mechanisms of enzyme activity and transport.

Processes involved in regulation of plasma amino acid concentrations are outlined in a general way in Figure 5-1. Regulation begins in the gastrointestinal tract with control of rate of passage of amino acids from stomach into intestine. After being absorbed, amino acids pass to the liver, an important site of protein synthesis and the major site of amino acid degradation. From the liver, amino acids are transported to other visceral organs and to the peripheral tissues. There the majority of protein synthesis occurs, but degradation of indispensable amino acids, other than the branched-chain group, is negligible. Control of amino acid

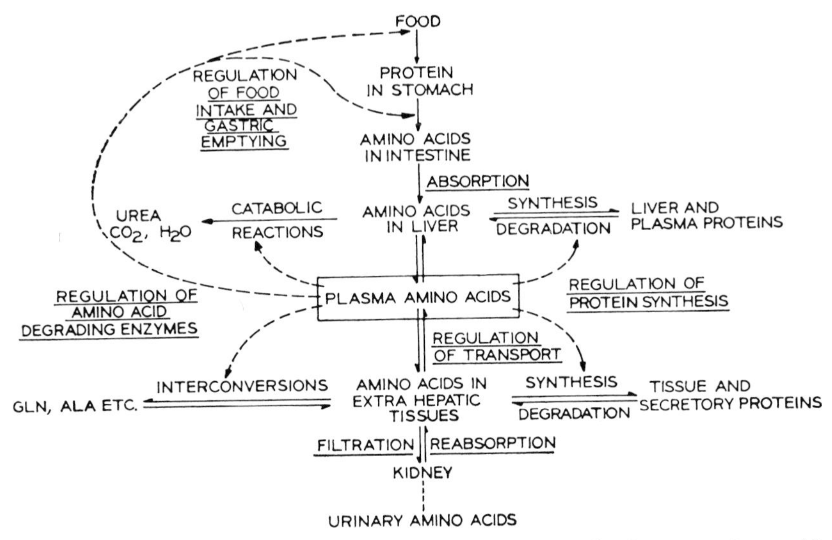

Figure 5-1. Schematic representation of regulation of plasma amino acid concentrations.

metabolism and hence, of body fluid amino acid concentrations, could be exerted through regulation of several mechanisms. These include regulation of gastric emptying, as well as transport during absorption, entry into tissues, and during reabsorption in the kidney. Others are regulation of incorporation of amino acids into proteins and regulation of degradation of amino acids, mainly in liver. Control of protein or amino acid intake through regulation of food intake remains a final possibility.

Storage

Storage in forms that can be mobilized during periods of deprivation is an important mechanism for regulation of plasma concentrations of energy substrates such as glucose and fatty acids. Evidence that amino acids are not stored comes from studies in which all of the essential nutrients except for one amino acid have been fed to animals at one time with the deleted amino acid being provided later. (61) Growth and tissue synthesis are curtailed even by a delay of a few hours in the provision of the missing amino acid. The period of delay tolerated differs with the amino acid. When lysine was provided twice daily to rats fed a lysine-deficient diet *ad libitum,* rats grew reasonably well. In contrast, when isoleucine was provided twice daily with an isoleucine-deficient diet fed ad libitum, they grew little better than rats fed only the isoleucine-deficient diet. (62) Little reserve is provided by the small amounts of free amino acids that accumulate in organ and tissue pools in relation to amino acid requirements (Table 5-VIII). Also, even though a low protein intake leads to depletion of body protein, a high protein intake results in very little increase in body protein content above that observed with a barely adequate diet. (63) These data show that surpluses of amino acids are not stored as protein.

The animal body thus cannot depend upon amino acid storage as a regulatory mechanism for disposal of surplus amino acids, nor as a mechanism for meeting the contingency of amino acid deprivation.

Although amino acids are not stored, many body proteins are degraded during periods of protein deprivation to provide amino acids for the maintenance of essential structures. Rapidly turning-over proteins, particularly in intestinal mucosa and liver are depleted rapidly when amino acid supply is inadequate and more slowly turning-over proteins, particularly those of muscle, are depleted later. This loss of functional protein may represent selective loss depending upon the state of the organism. Some proteins depleted from livers of animals fed a protein-free diet are not depleted from livers of starved animals even though both groups are protein-deficient. Evidently mechanisms are present for conservation of amino acid-degrading enzymes when primary need of the organims is energy, as in starvation. Conversely, mechanisms also exist for disposal of these enzymes, which fall to very low activities in protein depleted animals, when primary need is conservation of amino acids. (64) This homeostatic mechanism has great survival value for the organism with an inadequate supply of protein—a mechanism that permits redistribution of amino acids. It contributes to maintenance of plasma amino acid concentrations during periods of deprivation but not to removal of surpluses during periods of surfeit.

The Gastrointestinal Tract

The gastrointestinal tract is the initial site of regulation of the metabolism of nutrients. This organ system has evolved as a temporary reservoir from which the flow of nutrients to sites of metabolism is controlled, primarily through regulation of gastric emptying. (65,66) Among many factors that influence gastric emptying is the protein content of ingested food. Delayed emptying of protein reduces the probability of overloading the intestine. Evidence that digestive capacity of the small intestine is rarely exceeded comes from observations that most food proteins are completely digested.

The intestine has evolved as a mechanism with tremendous capacity, not only for digestion, but also for efficient trapping of essential nutrients. Unless rate of flow through the intestine is greatly accelerated owing to some abnormality, even toxic quantities of individual amino acids are almost completely absorbed. Tannous (67) observed that the rat excreted in the feces less than 1 percent of the radioactivity from a meal containing 5 percent of L-leucine-^{14}C (Figure 5-2). This observation is by itself evidence that there is no regulation of amino acid metabolism nor of plasma amino acid concentrations at sites of intestinal absorption.

The gastrointestinal tract as a whole functions efficiently as a mechanism for the conservation of ingested amino acids and even for the recovery of amino acids from the proteins of digestive secretions. (66) Regulation of gastric emptying becomes an important mechanism for ensuring that the flow of amino acids to sites of protein synthesis will be gradual and prolonged. The mechanism also provides that oxidative losses will be low unless the supply of amino acids is excessive. This regulation is needed not only because of rapid and extensive absorption of amino acids from the intestine, but also because of rapid degradation of amino acids when they build up to high concentrations in body fluids.

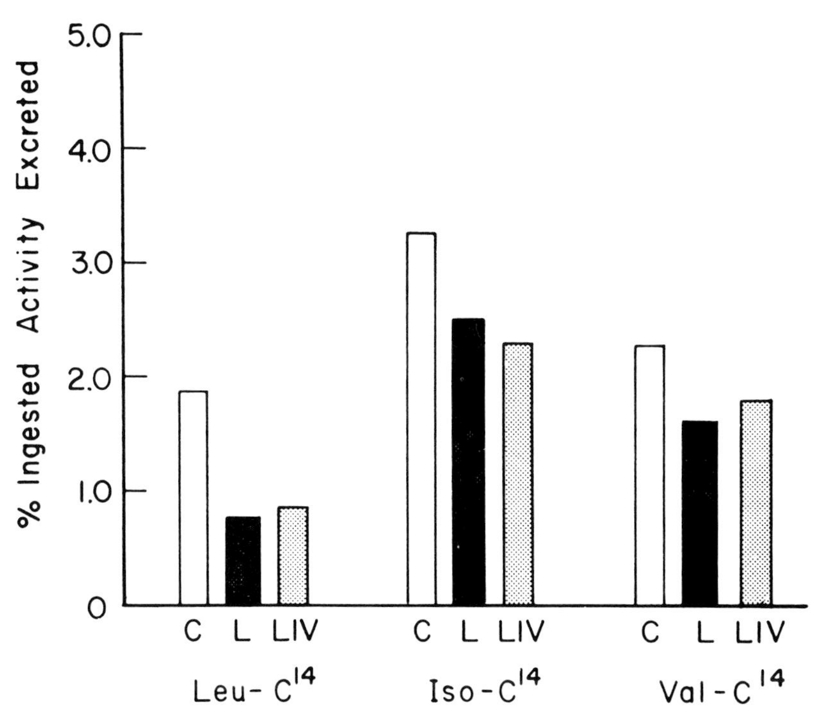

Figure 5-2. Fecal excretion of radioactivity by rats fed on diets containing isotopically-labeled leucine, isoleucine or valine. All rats were fed 10 gm of labeled diet on day one and unlabeled diet thereafter. Feces were collected for 4 days.[67]

C = 9% casin control diet.
L = C + 5% L-leucine.
LIV = L + 0.16% L-isoleucine + 0.15% L-valine.

Transport of Amino Acids into Cells and Reabsorption in the Kidney

Amino acids pass rapidly from the intestine into portal blood which flows to the liver. Thereafter, they are transported to other organs and tissues. Concentrations of amino acids in tissues ordinarily exceed those in blood, indicating an active transport of amino acids across cell membranes. The flux of a solute transported by a carrier-mediated system across a membrane is a function of the concentration of the solute in the medium up to the point of saturation of the carrier. When concentration of an amino acid in blood increases after a meal, as it does if concentration of the amino acid in the diet is increased enough, (68) its rate of entry into tissues should also increase.

Elevation of plasma glucose or amino acids stimulates insulin secretion. This in turn stimulates amino acid uptake by muscle and amino acid incorporation into muscle proteins. (69) Ingestion even of a protein-free diet will result in uptake of amino acids from blood plasma into muscle. (45) Glucagon stimulates uptake of amino acids by the isolated perfused liver (70) and by liver slices. (71) Amino acid uptake is also elevated in liver slices from rats fed a high protein diet. (72) There is evidence that glucagon or a high protein intake will elevate cyclic-3′,5′adenosine monophosphate (c-AMP) concentration in liver (73) and also that c-AMP stimulates amino acid uptake by liver slices. (71) These relationships suggest the following. An elevated protein intake stimulates insulin and glucagon secretion. Insulin stimulates uptake of amino acids into muscle and their incorporation into muscle proteins. This incorporation would occur even with a low protein intake provided the diet contained sufficient glucose. This action of insulin leads to utilization of amino acids preferentially for tissue protein synthesis. Glucagon stimulates uptake of amino acids by liver, especially when amino acid supply is large. Hence, glucagon would increase rate of removal of surpluses and lead to greater utilization of amino acids for gluconeogenesis. Regulation of transport of amino acids into tissues by these processes can contribute effectively on the one hand to conservation of amino acids for tissue protein synthesis, especially when the supply is limited. Conversely these processes may aid in removal of surpluses when supply is abundant and, hence, to stability of amino acid concentrations in plasma.

Reabsorption of amino acids by the kidney tubule is highly efficient as evidenced by low loss of amino acids in urine of animals fed a high protein diet. (74) Even when circulating concentrations of amino acids are in the range that leads to

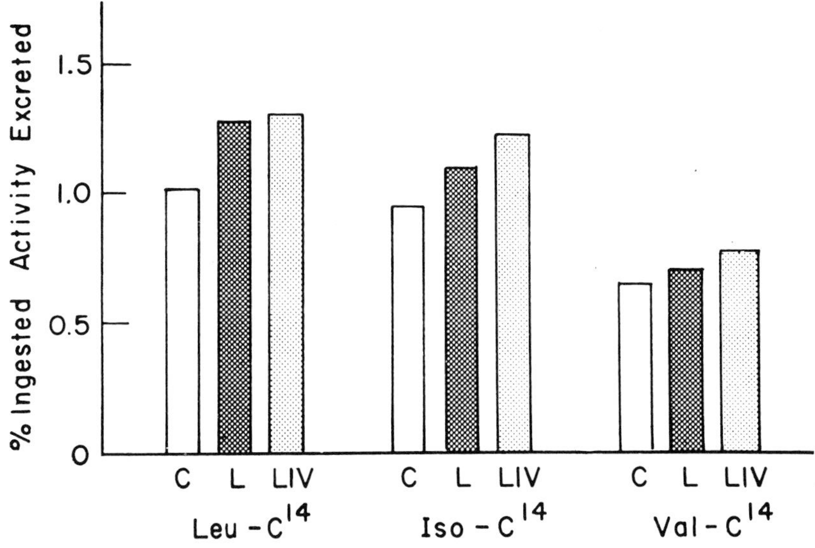

Figure 5-3. Urinary excretion of radioactivity by rats fed on diets containing isotopically-labeled leucine, isoleucine, or valine. All rats were fed 10 gm of labeled diet. Urine was collected for 24 hr.[67]
C = 9% casein control diet.
L = C + 5% L-leucine.
LIV = L + 0.16% L-isoleucine + 0.15% L-valine.

adverse effects, urinary loss is small. Figure 5-3 indicates the small amount of radioactivity (1%) excreted by rats fed a high leucine diet containing ^{14}C-leucine. (67) When rats are fed a diet containing 5 percent of tyrosine plasma tyrosine remains greatly elevated and the animals become severely debilitated even though tyrosine is excreted in the urine. Hence, urinary excretion should probably be viewed as a manifestation of toxicity rather than as a regulatory mechanism for homeostasis. (75) The kidney, like the intestine, conserves amino acids very efficiently. On the other hand, the kidney does not effectively regulate removal of excess circulating amino acids even when the circulating concentrations are in the toxic range. In fact, when the diet contains excess amino acids, food intake usually falls, indicating that curtailment of intake is an early regulatory response under these conditions rather than increased excretion.

Intravenous alimentation by-passes regulation of flow of nutrients to sites of absorption through control of stomach-emptying. Also, food intake regulating centers cannot curtail an excessive influx of amino acids. Regulation of plasma and body fluid amino acid concentrations then depends upon rate of utilization of amino acids for tissue synthesis and rate of removal by degrading systems. Influx of amino acids intravenously must, therefore, be controlled directly at a rate compatible with efficient tissue synthesis without overloading. Otherwise amino acids will be diverted into degradative pathways for energy or will accumulate in body fluids. If kidney or liver function is impaired it becomes especially critical not to overload systems for amino acid degradation. Amino acids themselves or their metabolic products will accumulate. Overloading severely malnourished infants with protein can cause deterioration of their condition and even lead to fatal consequences. (35)

Protein Synthesis

Regulation of protein synthesis by amino acid supply has been extensively studied. Munro and associates (76) and Sidransky and associates (77) have shown that ingestion of a complete mixture of amino acids initiates formation of polysomes, basic units involved in protein synthesis. The lack of one amino acid, tryptophan, prevents polysome formation. Jefferson and Korner (78) have shown that capacity of ribosomes from isolated perfused livers to synthesize protein increases as concentrations of amino acids in the perfusing medium are increased. Observations from our laboratory (Soliman and Harper, unpublished) illustrate the efficiency of amino acid incorporation attained under favorable conditions (Table 5-IX). Rats, briefly protein depleted and then fed six meals (3 per day) of a complete diet containing just enough ^{14}C-labeled lysine to meet the lysine requirement, retained 80 percent of absorbed lysine at the end of the 48-hour feeding period.

TABLE 5-IX

Distribution of Radioactivity in Rat Tissues at Intervals after Feeding Lysine-U-^{14}C

Hours after feeding	*Gut*	*Liver*	*Plasma*	*Total in body*
		% of administered dose		
0[1]	13.6	7.9	5.3	80.4[2]
24	9.4	6.0	3.7	71.5
48	8.0	5.3	2.6	58.9

[1] Zero time = 9 hr after the last six meals containing lys-U-^{14}C fed during the preceding 48 hr.

[2] Approximately 10% of the radioactivity was in body lipids.

The major fate of amino acids from diets containing moderate amounts of well-balanced proteins is incorporation into tissue proteins. If quantities of amino acids ingested by human subjects are not greatly in excess of requirements, and intake is appropriately proportioned over the day, plasma concentrations of indispensable amino acids do not rise appreciably. (79) Plasma amino acid concentrations, however, increase substantially for some time when protein intake is greatly increased.

Amino Acid Degradation

Amino aid degradation with elimination of degradation products is the major mechanism for disposal of surpluses of amino acids and control of plasma amino acid concentrations. This is true for two reasons.

1. Intestinal absorption and kidney reabsorption conserve amino acids, even when supply greatly exceeds needs for protein synthesis.
2. Tissue proteins do not accumulate nor are amino acids stored when surpluses of amino acids are ingested.

The first day after protein content of diet of young rats was increased from 5 percent to 50 percent, even though their total food intake was low, plasma amino acid concentration doubled (Figure 5-4). But, within a short time animals adapted to high protein intake and plasma amino acid concentrations, except for branched-chain amino acids, were not greatly elevated. (80) Initially the adaptation involved curtailment of food intake. Food intake subsequently

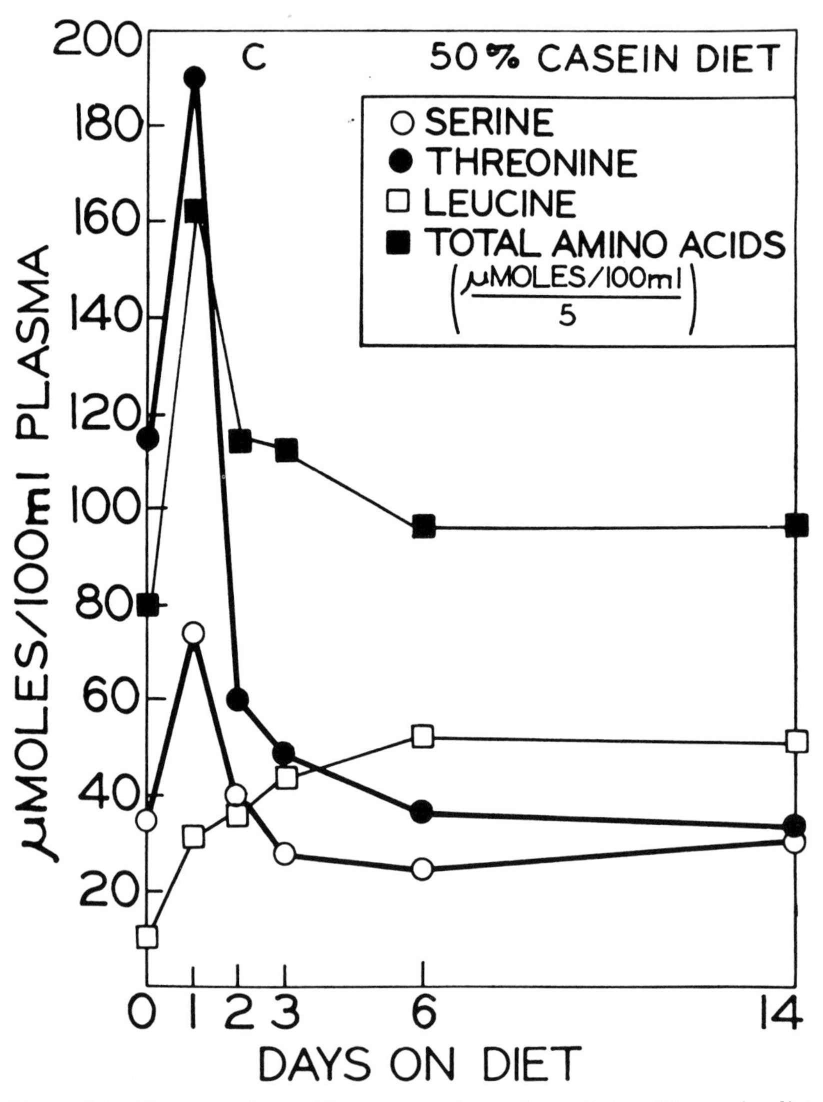

Figure 5-4. Plasma amino acid concentrations of rats fed a 5% casein diet for 10 days, then a diet containing 50% of casein.[80]

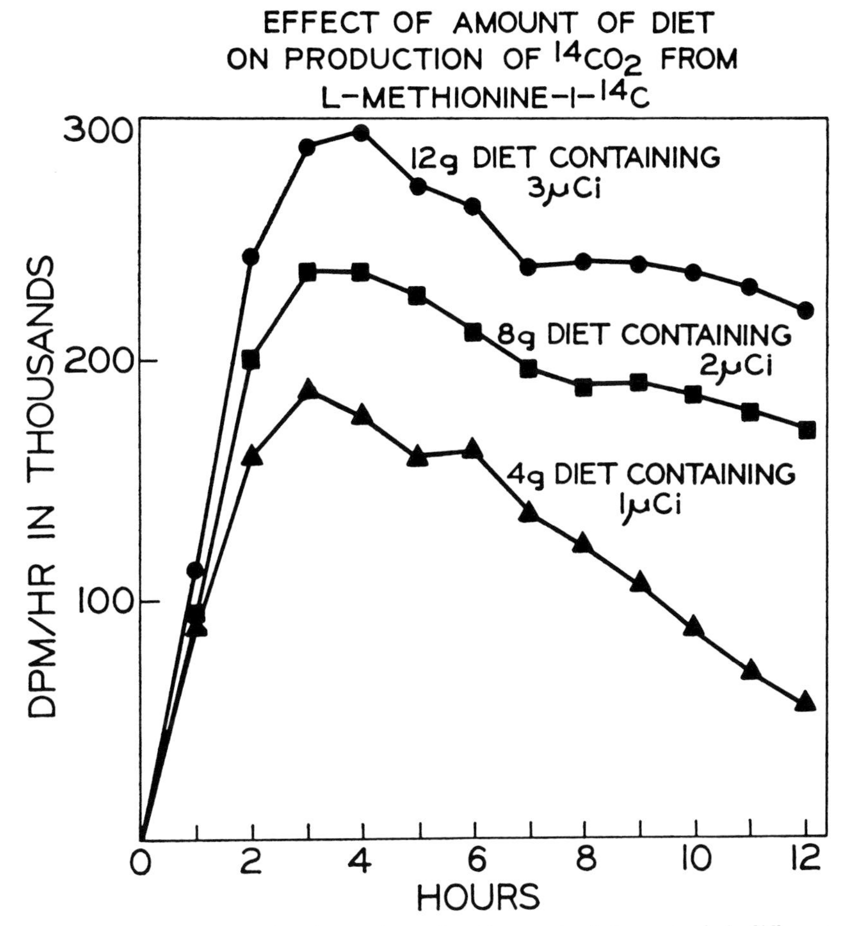

Figure 5-5. Radioactivity in carbon dioxide expired by rats fed different amounts of diet containing 10% casein + 3% L-methionine-1-^{14}C.[82]

returned almost to normal. Protein intake, nevertheless, was greatly elevated, indicating that metabolic adjustments had occurred to enable animals to dispose of surplus amino acids.

Two types of responses of amino acid degrading enzymes are involved in adjustment of the body to an increased amino acid load. The first might be considered a normal regulatory response. Ordinarily the body's metabolic machinery is not functioning at capacity; therefore, an influx of substrate, as an amino acid, results in increased activity of existing amounts of each enzyme involved in its metabolism. Kim and Miller (81) using isolated perfused liver, have shown that without change in tyrosine aminotransferase activity, tyrosine degradation increased substantially as tyrosine concentration of perfusate was increased. A similar response occurs with intact rats *in vivo* (Figure 5-5). In rats fed a high methionine diet, as size of the meal was increased, amount of expired $^{14}CO_2$ from ingested methionine-1-^{14}C increased. (82)

Despite increased rate of degradation in animals fed large amounts of individual amino acids, when tissue concentrations increase, plasma concentration of the amino acid in excess remains greatly elevated. This is particularly true when a load of a single amino acid is administered to an animal previously fed a low protein diet. (83) It is evidence that amino acid-degrading capacity has been exceeded even though enzyme activity may increase in response to increased substrate concentration.

In animals with a high intake of balanced protein, many enzymatic adaptations occur. Amounts of most amino acid-degrading enzymes increase. As a result, capacity of the body for amino acid degradation also increases. Activities of all of the urea cycle enzymes increase as protein intake is increased. (84) Histidase, (85) several amino-transferases, (86) and serine-threonine dehydratase (87) are examples of other enzymes that respond similarly. Figure 5-6 illustrates the response of histidase to increasing protein intake and the time course of the response in rats fed a high protein diet. (85)

Increased capacity for amino acid degradation owing to adaptations of this type is illustrated in Table 5-X. Blood histidine concentration and amount of $^{14}CO_2$ expired by rats given a meal containing a loading dose of ^{14}C-labeled histidine were measured at intervals after the meal. (101) Amounts of $^{14}CO_2$ excreted by rats previously fed a high protein diet or treated with glucagon and cortisol to induce histidine-degrading enzymes were well above those for rats fed a low protein diet. Plasma histidine concentration of those fed the low protein diet remained greatly elevated 12

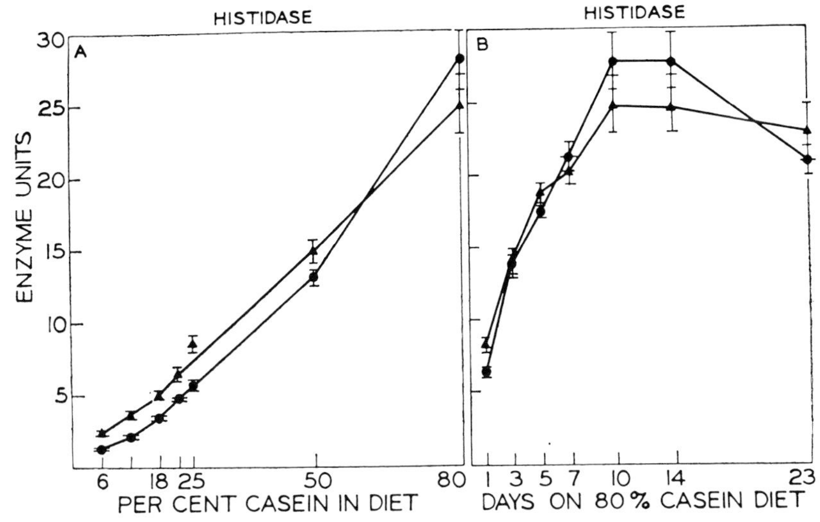

Figure 5-6. Effect of protein content of diet on liver histidase activity of rats (A) and change in histidase activity with time in rats fed 80% caesin diet (B).[85]

hours after the high histidine meal. In contrast, plasma histidine concentrations in those treated to induce the enzymes had returned nearly to the fasting value after 6 to 9 hours.

In sum, a change of intake of protein, and hence of amino acids, activates regulatory systems controlling amino acid-degrading capacity. This is an effective homeostatic mech-

TABLE 5-X

Relationship Between Activities of Histidine-Degrading Enzymes and Histidine-Degrading Capacity of the Rat

Treatment	*Enzyme Activity* *Histidase*	*Histidine aminotransferase*	$^{14}CO_2$ *from Histidine-U-*^{14}C *in 12 hr*	*Plasma histidine concentration at 12 hr*
	units/mg liver protein		*%*	*μmoles/100 ml*
Low protein (9% casein)	0.4	1.2	4.7	613
High protein (80% casein)	1.0	1.0	25.5	45
Low protein + glucagon + cortisol	1.0	22.3	28.6	21

Rats were administered 750 mg of histidine labeled with ^{14}C in a meal after they had received the treatments indicated for 10 days. (101).

anism. It should contribute toward conservation of amino acids for protein synthesis when degradative capacity falls in response to a low protein intake. Conversely, it should facilitate removal of amino acids when degradative capacity rises in response to a high protein intake.

The following types of evidence indicate that amino acids in the plasma pool must turn over rapidly:

1. Observation on amino acid intake in relation to the size of the plasma pool.
2. Observations on changes in plasma amino acids after a meal.
3. Observations on adaptive responses of animals to amino acid loads.

This is borne out by direct studies of amino acid turnover. Haider and Tarver (88) have shown that half of radioactivity in liver of rats administered ^{14}C-labeled lysine disappeared within 10 minutes. Neuberger and associates (89) reported similar observations for glycine in both plasma and liver although rate was somewhat slower. They found, however, that glycine entering muscle did not turn over rapidly, indicating that at least some free amino acids in muscle do not exchange readily with those in plasma. Our observations have shown that ^{14}C-lysine did not disappear from rat muscle after feeding ^{14}C-lysine, then a high unlabeled lysine diet, even though most of the unlabeled lysine that was ingested was oxidized (Soliman and Harper, unpublished). Despite this ability of the muscle to retain amino acids, turnover of individual amino acids in plasma is measured in minutes. Black and associates (90) obtained values of less than 15 minutes for the half-lives of several ^{14}C-labeled amino acids in cow plasma (Fig. 5-7). This is further evidence that both protein-synthesizing and amino acid-degrading systems, mainly responsible for constancy of plasma amino acid concentrations, function very efficiently in healthy, well-nourished animals.

ABNORMAL PLASMA AMINO ACID CONCENTRATIONS AND FOOD INTAKE REGULATION

An amino acid load is cleared rapidly from the blood of healthy, well-nourished animals or subjects and adaptation to continuous amino acid loading is rapid and effective.

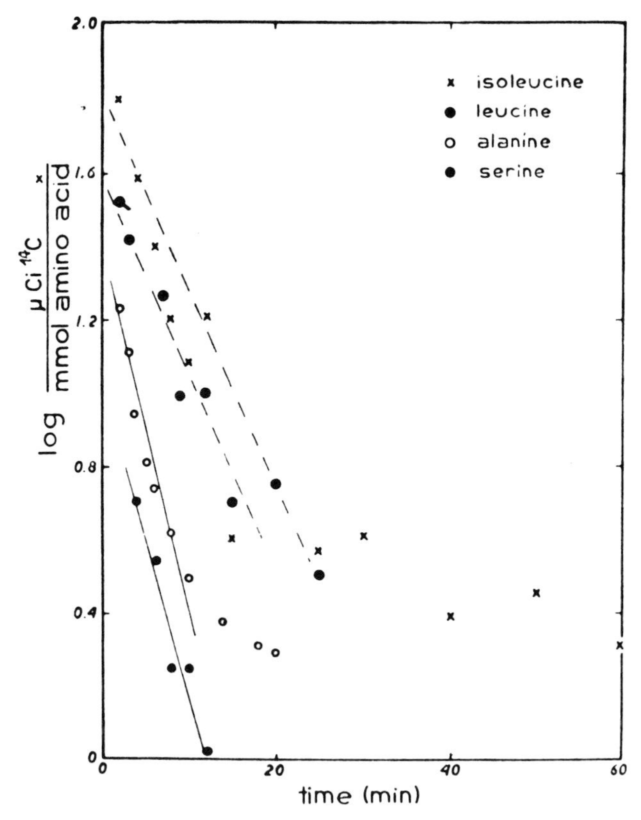

Figure 5-7. Changes in specific activity of plasma amino acids with time after intravenous injection of dairy cows with uniformly ^{14}C-labeled L-amino acids.[90]

These mechanisms on the other hand are less efficient in the malnourished organism. If dietary protein intake is low and dietary amino acid pattern unbalanced, plasma and body fluid amino acid patterns may deviate from the standard state. Clearance of amino acids may also be slow. Values for rats fed low protein diets shown in Figure 5-8, were from a study by Perez and Harper (unpublished). Amino acid concentrations were measured at intervals in plasma from rats that had been fed a small meal of either a low protein control diet or the same diet unbalanced by addition of an amino acid mixture devoid of histidine. The horizontal line represents the

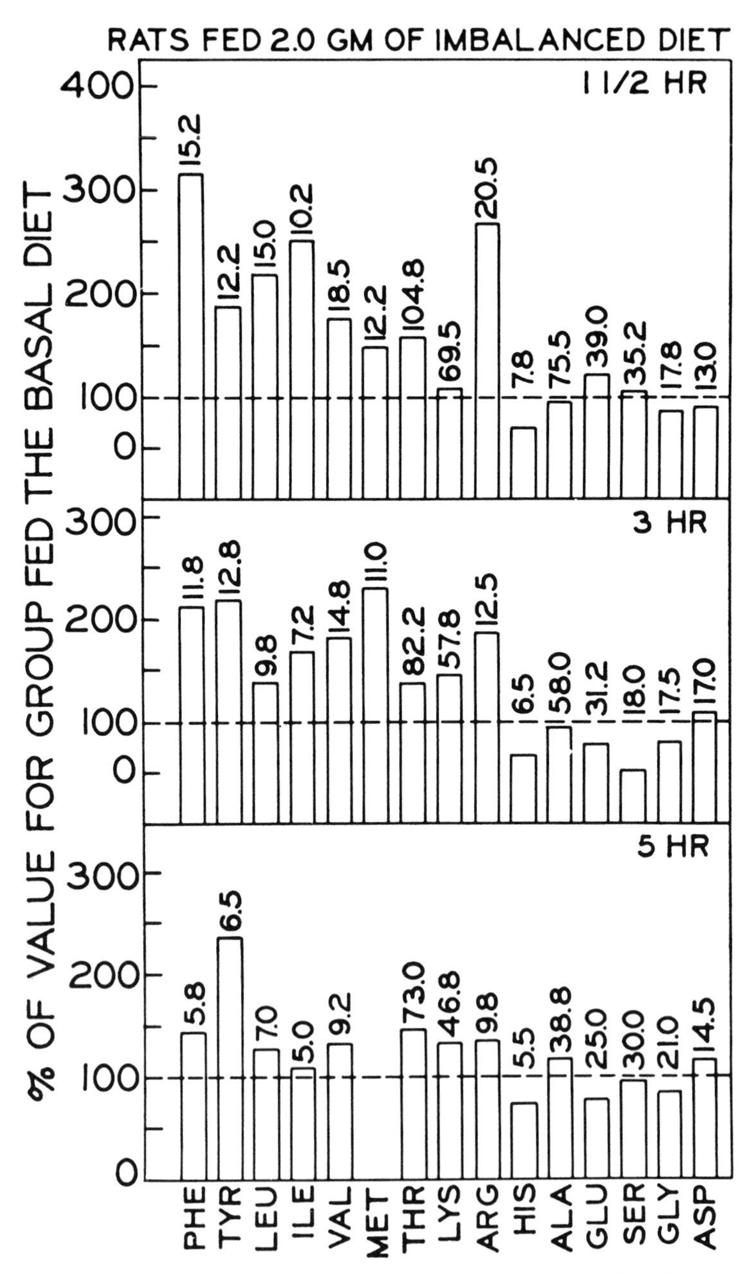

Figure 5-8. Changes in plasma amino acid concentrations with time in rats fed 2.0 gm of diet containing 6% caesin + 6% of amino acid mixture devoid of histidine. Dotted line indicated control values set at 100%. Histidine intakes of both groups were the same (Perez and Harper, unpublished).

control value as 100 percent. It is not evident, but the control values fluctuated only within narrow limits and actually tended to fall with time rather than rise. On the other hand, amino acids added to the diet to unbalance the amino acid pattern tended to accumulate in plasma and to be cleared slowly; while histidine, the amino acid not added, fell well below the control value even though histidine intakes of the two groups were the same. A comparable study in which an amino acid mixture devoid of threonine was used (91) indicated that the pattern in muscle resembled that in plasma. Liver was less consistently affected. Brain amino acid pattern was also altered in rats fed a diet in which there was an imbalance of amino acids. (92)

Accumulation in plasma of the amino acids added to a diet to create an imbalance is associated with low food intake, a slow rate of growth and low activities of several of the enzymes of amino acid degradation. (93) Thus, with a low intake of protein, (a) amino acid degrading capacity is low. With an unbalanced dietary amino acid pattern as well, (b) a substantial portion of indispensable amino acids consumed cannot be used for protein synthesis. These two factors probably account for the unbalanced plasma amino acid pattern. (94)

Observations of McLaughlan and associates (68) (Fig. 5-9) show the effect of altering the concentration of one amino acid in the diet over the range from deficient to excessive. When intake of each amino acid tested was at about the requirement level, plasma concentration remained close to the fasting value, indicating that intake and utilization were in balance. An intake below the requirement resulted in a plasma concentration below the fasting value and an intake above the requirement resulted in accumulation of the amino acid in plasma.

When the mechanisms for disposal of surpluses of amino acids are overloaded or have failed to respond, reducing amino acid intake is the only remaining way for the body to maintain homeostasis of plasma and body fluid amino acid concentrations. This might well be considered as a facet of

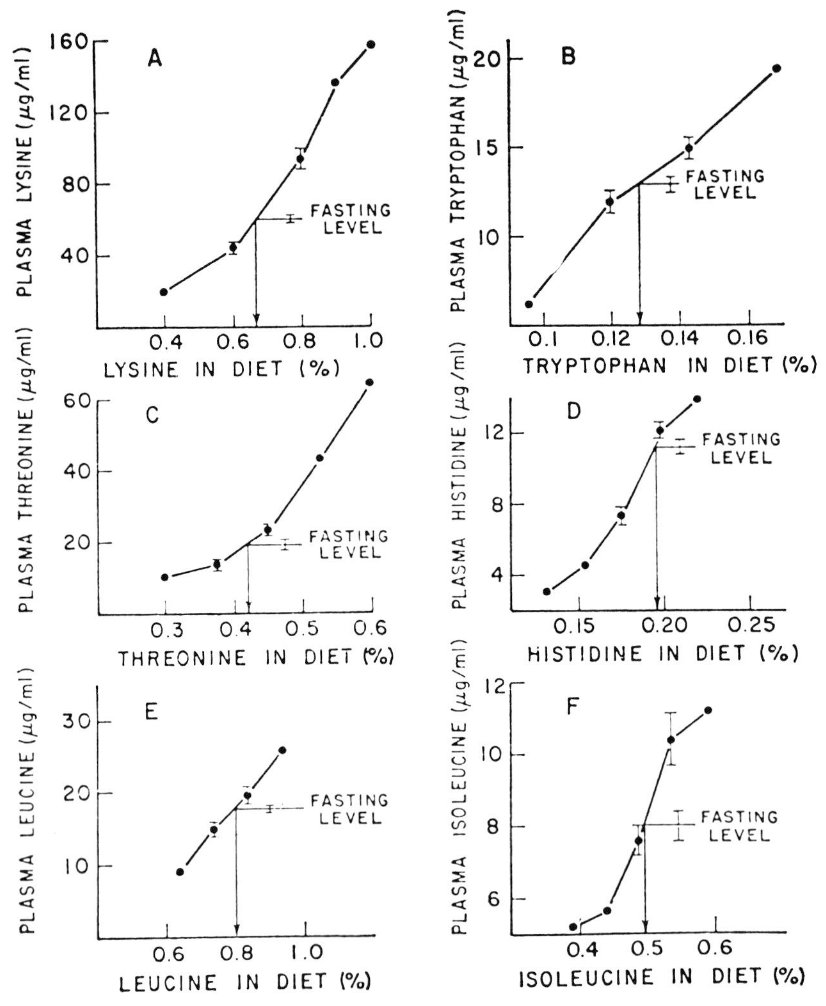

Figure 5-9. Plasma amino acid concentrations of rats fed graded increments of one indispensable amino in an otherwise complete diet.[68]

regulation of transport, with the organism itself serving as a complex feedback mechanism. This mechanism is eliminated by intravenous alimentation, but in a number of situations with animals fed normally there is a close association between depressed food intake and greatly elevated plasma amino acid concentrations or deviations in plasma amino acid pattern. (94) Food intake depression is observed when animals are fed diets with amino acid imbalances and the plasma pattern resembles that shown in Figure 5-8. The situation is similar when the diet is deficient in an amino acid. An excess of protein or of an individual amino acid depresses food intake and this is accompanied by elevated plasma amino acid concentrations.

In a study (95) in which rats were allowed to eat a low protein control diet freely and were infused continuously with an unbalanced amino acid mixture, their food intake fell sharply. This response did not reduce the intravenous inflow of the unbalanced amino acid mixture. With prolonged infusion of a large amount of the mixture some animals became ill and some died. Also, if protein intake is low and only one or two amino acids are ingested in great excess, the enzymatic responses are slow and small, and food intake may remain depressed. If food intake depression is great, energy needs cannot be met, and if the amino acid intake remains high despite depressed food intake, the animal continues to deteriorate. (94)

These observations suggest that food intake depression is a protective mechanism. It comes into play when limits of other homeostatic mechanisms for regulating plasma and body fluid amino acid concentrations, such as protein synthesis and amino acid degradation, are exceeded. Survival value of depressed food intake is illustrated by observations on animals fed a high methionine diet in a cold environment. Here food intake stimulation, owing to need for calories, overrides food intake-depressing effect of excess methionine and signs of methionine toxicity become more severe. (96) Also, as mentioned above, animals deteriorate when infused intravenously with an unbalanced mixture of amino acids, so depressed food intake does not reduce amino acid inflow. They do not when the mixture is fed as part of the diet and intake is reduced. (97) Several reports, especially from Sidransky and associates, indicate that rats force-fed amino acid deficient diets do not survive as long as controls fed *ad libitum* but suffering severe food intake depression. (94) These observations emphasize that depressed food intake is not necessarily an adverse effect. It can be a response with great survival value and an indicator that the capacity of the body to maintain homeostasis has been exceeded as a result of ingestion of the particular diet. Leung and Rogers (98) have identified an area in the brain involved in food intake regulation that is responsive to alterations in plasma amino acid pattern.

The degree of abnormality of plasma amino acid pattern tolerated by animals before food intake is depressed is not clearly established. It appears to depend upon duration of the abnormal pattern after a meal, which in turn depends upon capacity of the body to degrade amino acids. Rats maintained at room temperature gradually adapt when fed a diet with an amino acid imbalance.

Their food intake increases even though the plasma amino acid pattern remains abnormal. This adaptation is accompanied by a gradual, small increase in threonine dehydratase activity representative of the responses of amino acid-degrading enzymes. (99) Rats exposed to cold adapt similarly, but more rapidly and here too the plasma amino acid pattern remains abnormal. (100) These observations suggest that, to satisfy its need for energy, an animal will eat a diet that causes the plasma amino acid pattern to become abnormal until it can tolerate no greater change in plasma amino acid concentrations. Whether it can meet its need for energy before this point will depend upon its capacity to dispose of amino acids and amino acid composition of the diet.

This implies that energy need rather than amino acid pattern is the major determinant of food intake. If, on the other hand, amino acids are ingested in excess of capacity of the organism to remove them, food, and hence energy intake, will be depressed as plasma amino acid concentrations deviate too greatly from the standard state.

HOMEOSTASIS OF PLASMA AMINO ACID CONCENTRATION—RECAPITULATION

Control of plasma amino acid concentrations depends upon regulatory or feedback systems of the type portrayed at the top of Figure 5-10. These systems include (1) a means of monitoring existing concentrations at any time, (2) a means of detecting deviations in existing concentrations from the standard state, and (3) mechanisms to do the work necessary to correct deviations. The lower part of Figure 5-10 represents an attempt to portray as feedback systems, part of the complex of regulatory mechanisms indicated in Figure 5-1. Each segment contributes to maintenance of homeostasis of plasma amino acid concentrations in the mammalian body. It is obviously an oversimplification, but application of it to a situation in which amino acids are ingested in considerable excess recapitulates, in brief, much of the previous discussion.

Initially, influx of amino acids into blood activates mechanisms that regulate transport rate of amino acids into cells,

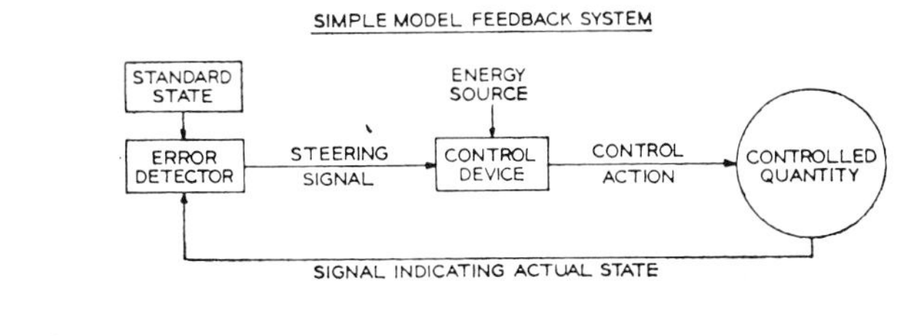

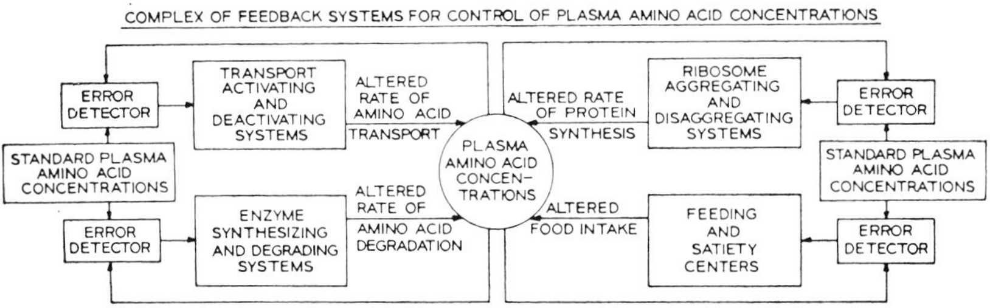

Figure 5-10. Simple model feedback system (above). Schematic representation of feedback systems for control of plasma amino acid concentrations (below).

especially muscle. As transport increases rate of removal of amino acids from plasma increases. Influx of amino acids into cells activates mechanisms that regulate protein synthesis, such as polysome formation. This increases rate of protein synthesis which, in turn, further increases removal rate of amino acids. Because capacity for protein synthesis is limited, amino acids in excess of immediate needs for tissue formation accumulate in plasma and other body fluids. If excess is not too great amino acid oxidation rate increases owing solely to increased substrate concentration. With ordinary protein intakes in healthy subjects this response is adequate to prevent an inordinate rise in plasma amino acid concentrations. If, however, degrading systems are inadequate or load of amino acids too great, plasma amino acid concentrations remain elevated or abnormal in pattern and initiate a signal to centers that regulate food intake. Food intake depression decreases influx of amino acids into blood. In the animal receiving a surplus of well-balanced protein, elevated plasma amino acid concentrations also activate mechanisms that both regulate enzyme synthesis and degradation and initiate induction of amino acid-degrading enzymes. Resulting increases in amino acid-degrading capacity increase potential for removal of excess amino acids and for reducing time required to restore plasma amino acid concentrations to the standard state. As clearance rate of blood amino acids increases food intake increases and homeostasis is attained again but with an elevated rate of amino acid metabolism.

If, owing to malnutrition, undernutrition, pathologic states, or genetic defects, regulatory systems are defective or incapable of responding to appropriate signals from surplus or deficit of amino acids plasma amino acid concentrations are not restored to the standard state or are restored only slowly. Depending on severity of the condition, the organism may gradually deteriorate unless some ameliorating measure is provided. For example, a diet low in an amino acid for which the subject lacks an important degrading enzyme owing to a genetic defect might be provided. Alternatively, the organism might function suboptimally if a state compatible with survival but not normal activity is achieved.

SUMMARY

Nitrogen requirements are specific for indispensable amino acids and nonspecific for additional nitrogen. Seventy mg nitrogen per kilogram (kg) of body weight (wt) per day is required for average adult men, based on balance studies in which proteins of high nutritional quality were fed. This exceeds measured average nitrogen losses of about 54 mg/kg of body wt/day, indicating that even high quality proteins are used with an efficiency of only about 75 percent when consumed in quantities that meet requirements.

Requirements for indispensable amino acids are low, although probably not as low as is generally accepted, because unmeasured nitrogen losses from sweat and other minor routes are not usually taken into account when these requirements are determined by the balance procedure. Nevertheless, the amount of nitrogen required from specific amino acids represents only 20 to 25 percent of total nitrogen.

Infants require about four times as much nitrogen per kilogram of body weight as adults, but about six to eight times as much of each indispensable amino acid.

When nitrogen in the diet is from high quality proteins, as protein consumption decreases, total nitrogen deficit occurs before deficits of specific amino acids. When nitrogen in the diet is from poor quality proteins, specific amino acid deficits may occur when total nitrogen requirement has been met.

Efficiency of nitrogen utilization is influenced by total nitrogen intake, dietary amino acid pattern, energy intake, sweat loss owing to high environmental temperatures or vigorous activity, and by infections, trauma, and stress.

Plasma amino acid concentrations in healthy individuals consuming a nutritionally well-balanced diet fluctuate relatively little. They fall only slightly after prolonged fasting and are restored rapidly to fasting levels after an amino acid load, indicating that they are well regulated.

Regulatory systems that contribute to the stability of plasma amino acid concentrations include regulation of (1) intake, (2) gastric emptying, (3) transport into cells, (4) protein synthesis and (5) amino acid degradation. Amino acid absorption from the intestine and reabsorption by the kidney are conservation and trapping systems only. Amino acids are not stored, although tissue proteins may be degraded and the amino acids redistributed during periods of deprivation.

If regulatory systems are deficient or defective owing to malnutrition, disease, or genetic defects, or are overloaded, plasma amino acid patterns and concentrations may become abnormal. If the organism is healthy and well-nourished adaptations to an overload occur within a short time and abnormalities are only transitory.

REFERENCES

1. Williams, H. H., Harper, A. E., Hegsted, D. M., Arroyave, G., and Holt, L. E., Jr.: Nitrogen and amino acid requirements. In Harper, A. E., and Hegsted, D. M. (Eds.): *Evaluation of Protein Nutrition*. National Academy of Science–National Research Council, Washington, D.C., in press.
2. Food and Nutrition Board: *Evaluation of Protein Nutrition* Publication 711, National Academy of Science-National Council, Washington, D.C., 1959.
3. *FAO/WHO Energy and Protein Requirements*. WHO Technical Report Series No. 522. Geneva, 1973.
4. *FAO Report. Protein Requirements*. Nutrition Studies No. 16. FAO, Rome, 1957.
5. *WHO/FAO Report*. Protein Requirements. WHO Technical Report Series No. 301, WHO, Geneva, 1965.
6. Terroine, E. F. In Waterlow, J. C., and Stephen, J. M. F. (Eds.): *Human Protein Requirements and Their Fulfilment in Practice*. FAO/WHO/Josiah Macy Foundation, 1957. p. 16.
7. Lusk, G.: *The Science of Nutrition,* 3rd ed. Philadelphia, Saunders, 1923, pp. 334–346.
8. Wallace, W. M.: Nitrogen content of the body and its relation to retention and loss of nitrogen. *Fed Proc, 18:* 1125–1130, 1959.
9. Hegsted, D. M.: Variation in requirements of nutrients—amino acids. *Fed Proc, 22:* 1424–1430, 1963.

10. Block, R. J., and Mitchell, H. H.: The correlation of the amino acid composition of proteins with their nutritive value. *Nutr Abstr Rev, 16:* 249–278, 1946.
11. Bricker, M., Mitchell, H. H., and Kinsman, G. M.: The protein requirements of adult human subjects in terms of the protein contained in individual foods and food combinations. *J Nutr, 30:* 269–283, 1945.
12. Hegsted, D. M.: False estimates of adult requirements. *Nutr Rev, 10:* 257, 1952.
13. Young, V. R., and Scrimshaw, N. S.: Endogenous nitrogen metabolism and plasma free amino acids in young adults given a protein-free diet. *Br J Nutr, 22:* 9–20, 1968.
14. Calloway, D. H., and Margen, S.: Variation in endogenous nitrogen excretion and dietary nitrogen utilization as determinants of human protein requirements. *J Nutr, 101:* 205–216, 1971.
15. Sirbu, E. R., Margen, S., and Calloway, D. H.: Effect of reduced protein intake on nitrogen loss from human integument. *Am J Clin Nutr, 20:* 1158–1165, 1967.
16. Calloway, D. H., Odell, A. C. R., and Margen, S.: Sweat and miscellaneous nitrogen losses in human balance studies. *J Nutr, 101:* 775–786, 1971.
17. Almquist, H. J.: Principles of amino acid balance. *Arch Biochem Biophys, 46:* 250, 1953.
18. Harper, A. E.: Advances in our knowledge of protein and amino acid requirments. *Fed Proc, 18:* Supplment 3, 104-113, 1959.
19. Scrimshaw, N. S., Young, V. R., Schwartz, R. Piche, M. L., Das, J. B.: Minimum dietary essential amino acid-to-total nitrogen ratio for whole egg protein fed to young men. *J Nutr, 89:* 9–18, 1966.
20. Rose, W. C., and Wixom, R. L.: The amino acid requirements of man XVI. The role of the nitrogen intake. *J Biol Chem, 217:* 997–1104, 1955.
21. Rose, W. C.: The amino acid requirements of adult man. *Nutr Abstr Revs, 27:* 631–647, 1957.
22. Leverton, R. M.: Amino acid requirements of young adults. In Albanese, A. A. (Ed.): *Protein and Amino Acid Requirements.* New York, Academic Press, 1959, chap. 15.
23. Munro, H. N.: Amino acid requirements and metabolism and their relevance to parenteral nutrition. In Wilkinson, A. (Ed.): *Parenteral Nutrition.* In press.
24. Burrill, L. M., and Shuck, C.: Phenylalanine requirements with different levels of tyrosine. *J Nutr, 83:* 202–208, 1964.
25. Swendseid, M. E., Feeley, R. J., Harris, C. L., and Tuttle, S. G.: Egg protein as a source of essential amino acids. Requirements for young adults studied at two levels of nitrogen intake. *J Nutr, 68:* 203–211, 1959.

26. Swendseid, M. E., Harris, C. L., Tuttle, S. G.: The effect of sources of nonessential nitrogen on nitrogen balance in young adults. *J Nutr, 71:* 105–108, 1960.
27. Scrimshaw, N. S., Young, V. R., Schwartz, R., Piche, M. L., and Das, J. B.: Minimum dietary essential amino acid to total nitrogen ratio for whole egg protein fed to young men. *J Nutr, 89:* 9–18, 1966.
28. Huang, P. C., Young, V. R., Cholakos, B., and Scrimshaw, N. S.: Determination of the minimum dietary essential amino acid-to-total nitrogen ratio for beef protein fed to young men. *J Nutr, 90:* 416–422, 1966.
29. Scrimshaw, N. S., Young, V. R., Huang, P. C., Thanangkul, O., and Cholakos, B. V.: Partial dietary replacement of milk protein by nonspecific nitrogen in young men. *J Nutr, 98:* 9–17, 1969.
30. Snyderman, S. E., Holt, L., Emmett, Jr., Dancis, J., Roitman, E., Boyer, A., and Balis, M. E.: Unessential nitrogen: A limiting factor for human growth *J Nutr, 78:* 57–72, 1962.
31. Fisher, H., Brush, M. K., Griminger, P.: Reassessment of amino acid requirements of young women on low nitrogen diets. 1. Lysine and tryptophan. *Am J Clin Nutr, 22:* 1190–1196, 1969.
32. Hegsted, D. M.: Protein requirement in man. *Fed Proc, 18:* 1130–1136, 1959.
33. Hartsook, E. W., and Mitchell, H. H.: The effect of age on the protein and methionine requirements of the rat. *J Nutr, 60:* 173–196, 1956.
34. Foman, S. J., Zeigler, E. E., Thomas, L. N., and Filer, L. J.: Protein requirements of normal infants between 8 and 56 days of age. In Visser, H. K. A. (Ed.) *Metabolic Processes in the Foetus and Newborn Infant.* Nutricia Symposium, Vol. III. Stenfert Kroese, Leidein, 1971.
35. Holt, L. E., Jr., Gyorgy, P., Pratt, E. L., Snyderman, S. E., and Wallace, W. M.: *Protein and Amino Acid Requirements in Early Life.* New York, New York University Press, 1960.
36. FAO: Amino Acids Content of Foods and Biological Data on Proteins Rome: FAO, 1968.
37. Holmes, E. G.: An appraisal of the evidence upon which recently recommended protein allowances have been based. *World Rev Nutr Diet, 5:* 237–274, 1965.
38. Romo, G. S., and Linkswiler, H.: Effect of level and pattern of essential amino acids on nitrogen retention of adult man. *J Nutr, 97*: 147–153, 1969.
39. Clark, H. E., Neeney, M. A., Goodwin, A. F., Goyal, K., and Mertz, E. T.: Effect of certain factors on nitrogen retention and lysine requirements of adult human subjects. IV. Total nitrogen intake. *J Nutr, 81:* 223–229, 1963.

40. Clark, H. E., Fugate, K., and Allen, P. E.: Effect of four multiples of a basic mixture of essential amino acids on nitrogen retention of adult human subjects. *Am J Clin Nutr, 20:* 233–242, 1967.
41. Kies, C., Williams, E., and Fox, H. M.: Determination of first limiting nitrogenous factor in corn protein for nitrogen retention in human adults. *J Nutr, 86:* 350–356, 1965.
42. Kies, Constance, Fox, H. M., and Williams, E. R.: Effect of nonspecific nitrogen supplementation on minimum corn protein requirement and first-limiting amino acid for adult men. *J Nutr, 92:* 377–383, 1967.
43. Munaver, S. M., and Harper, A. E.: Amino acid balance and imbalance II. Dietary level of protein and lysine requirement. *J Nutr, 69:* 58–64, 1959.
44. Said, A. K., and Hegsted, D. M.: Evaluation of dietary protein quality in adult rats. *J Nutr, 99:* 474–480, 1969.
45. Munro, H. N.: General aspects of the regulation of protein metabolism by diet and by hormones. In Munro, H. N., and Allison, J. B. (Eds.): *Mammalian Protein Metabolism.* New York, Academic Press, 1964, vol. I, pp. 381–481.
46. Waterlow, J. C.: Observations on the mechanism of adaptation to low protein intakes. *Lancet, ii;* 1091–1097, 1968.
47. Harper, A. E.: Adaptability and amino acid requirements. In *Metabolic Adaptation and Nutrition.* Science Publication No. 222. Pan American Health Organization, Washington, D.C., 1971, pp. 8–20.
48. Munro, H. N.: Carbohydrate and fat as factors in protein utilization and metabolism. *Physiol Rev, 31:* 449–488, 1951.
49. Calloway, D. H., and Spector, H.: Nitrogen balance as related to caloric and protein intake in active young men. *Am J Clin Nutr, 2:* 405–412, 1954.
50. Clark, H. E., Yang, S. P., Reitz, L. L., and Mertz, E. T.: The effect of certain factors on nitrogen retention and lysine requirements of adult human subjects. I. Total caloric intake. *J Nutr, 72:* 87–92, 1960.
51. Rose, W. C., Coon, M. J., and Lambert, G. F.: The amino acid requirements of man. VI. The role of the calorie intake. *J Biol Chem, 210:* 331–342, 1954.
52. Anderson, H. L., Heindel, M. B., and Linkswiler, H.: Effect on nitrogen balance of adult man of varying source of nitrogen and level of calorie intake. *J Nutr, 99:* 82–90, 1969.
53. Border, J. R.: Metabolic response to short-term starvation, sepsis and trauma. *Surg Ann,* 1970, pp. 11–34.
54. Scrimshaw, N. S. Protein deficiency and infective disease. In Munro, H. N., and Allison, J. B. (Eds.): *Mammalian Protein Metabolism.* New York, Academic Press, 1964, vol. II, pp. 569–593.

55. Allison, J. B.: Biological evaluation of proteins. *Physiol Rev, 35:* 664–700, 1955.
56. Munro, H. N.: Free amino acid pools and their role in regulation. In *Mammalian Protein Metabolism.* New York, Academic Press, 1970, vol. IV, pp. 299–386.
57. Lang, K.: Physiology and metabolism of amino acids. In Meng, H. C., and Law, D. H. (Eds.) : *Parenteral Nutrition.* Springfield, Ill., Thomas, 1970, chap. 17.
58. Anderson, H. L., and Linkswiler, H.: Effect of source of dietary nitrogen on plasma concentrations and urinary excretion of amino acids of men. *J Nutr, 99:* 91–100, 1969.
59. Adibi, S. A.: Influence of dietary deprivations on plasma concentration of free amino acids of man. *J Appl Physiol, 25:* 52–57, 1968.
60. Coulson, R. A., and Hernandez, T.: Amino acid catabolism in the intact rat. *Am J Physiol, 215:* 741–746, 1968.
61. Elman, R.: Time factor in retention of nitrogen after intraveneous injection of a mixture of amino acids. *Proc Soc Exp Biol Med, 40:* 484–487, 1939.
62. Spolter, P. D., and Harper, A. E.: Utilization of injected and orally administered amino acids by the rat. *Proc Soc Exp Biol Med, 106:* 184–189, 1961.
63. Mayer, J., and Vitale, J. J.: Thermochemical efficiency of growth in rats. *Am J Physiol, 189:* 39–42, 1957.
64. Harper, A. E.: Effect of variations in protein intake on enzymes of amino acid metabolism. *Can J Biochem, 43:* 1589–1603, 1965.
65. Rogers, Q. R., and Harper, A. E.: Digestion of proteins: Transfer rates along the gastrointestinal tract. In Munro, H. N. (Ed.) : *The Role of the Gastrointestinal Tract in Protein Metabolism.* Oxford, Blackwell Scientific Publications, Ltd., 1964, p. 3–24.
66. Rogers, Q. R., and Harper, A. E.: Protein digestion: Nutritional and metabolic considerations. In Bourne, G. H. (Ed.) : *World Review of Nutrition and Dietetics.* New York, Hafner Publishing Company, Inc., 1966. p. 250–291.
67. Tannous, R. I.: Metabolic studies on leucine, isoleucine and valine antagonism in the rat. D. Sc. Thesis, Massachusetts Institute of Technology, Cambridge, 1963.
68. McLaughlan, J. M., and Illman, W. I.: Use of free plasma amino acid levels for estimating amino acid requirements of the growing rat. *J Nutr, 93:* 21–24, 1967.
69. Wool, I. G., Stirewalt, W. S., Kurhara, K., Low, B., Bailey, P., and Oyer, D.: Mode of action of insulin in the regulation of protein biosynthesis in muscle. In Astwood, E. B. (Ed.) : *Recent Progress in Hormone Research.* New York, Academic Press, 1968, p. 139–208.

70. Mallette, L. E., Exton, J. H., and Park, C. R.: Effects of glucagon on amino acid transport and utilization in the perfused rat liver. *J Biol Chem, 244:* 5724–5728, 1969.
71. Tews, J. K., Woodcock, N. A., and Harper A. E.: Stimulation of amino acid transport in rat liver slices by epinephrine, glucagon and adenosine 3′,5-monophosphate. *J Biol Chem, 245:* 3026–3032, 1970.
72. Tews, J. K., Woodcock, N. A., and Harper, A. E.: Effect of protein intake on amino acid transport and adenosine 3′-5′-monophosphate content in rat liver. *J Nutr, 102:* 409–418, 1972.
73. Jost, J.-P., Hsie, A., Hughes, S. D., and Ryan, L.: Role of cyclic adenosine 3′,5-monophosphate in the induction of hepatic enzymes. *J Biol Chem, 245:* 351–357, 1970.
74. Sauberlich, H. E., and Baumann, C. A.: The effect of dietary protein upon amino acid excretion by rats and mice. *J Biol Chem, 166:* 417–428, 1946.
75. Boctor, A. M.: Some nutritional and Biochemical Aspects of Tyrosine Toxicity and Lysine Availability. Ph.D. Thesis, Massachusetts Institute of Technology, Cambridge, 1967.
76. Munro, H. N. Role of amino acid supply in regulating ribosome function. *Fed Proc, 27:* 1231–1237, 1968.
77. Sidransky, H., Sarma, D. S. R., Bongiorno, M., and Verney, E.: Effect of dietary tryptophan on hepatic polyribosomes and protein synthesis in fasted mice. *J Biol Chem 243:* 1123–1132, 1968.
78. Jefferson, L. S., and Korner, A.: Influence of amino acid supply on ribosomes and protein synthesis of perfused rat liver. *Biochem J, 111:* 703–712, 1969.
79. Anderson, H. L., and Linkswiler, H.: Effect of source of dietary nitrogen on plasma concentrations and urinary excretion of amino acids of men. *J Nutr, 99:* 91–100, 1969.
80. Anderson, H. L., Benevenga, N. J., and Harper, A. E.: Associations among food and protein intake, serine dehydratase and plasma amino acids. *Am J Physiol, 214:* 1008–1013, 1968.
81. Kim, J. H., and Miller, L. L.: The functional significance of changes in activity of the enzymes, tryptophan pyrrolase and tyrosine transaminase, after induction in intact rats and in the isolated, perfused rat liver. *J Biol Chem, 244:* 1410–1416, 1969.
82. Benevenga, N. J., and Harper, A. E.: Effects of glycine and serine on methionine metabolism in rats fed diets high in methionine. *J Nutr, 100:* 1204–1214, 1970.
83. Sauberlich, H. E.: Studies on the toxicity and antagonism of amino acids for weanling rats. *J Nutr, 75:* 61–72, 1961.
84. Schimke, R. T.: Studies on factors affecting the levels of urea cycle enzymes in rat liver. *J Biol Chem, 238:* 1012–1018, 1963.

85. Schirmer, M. D., and Harper, A. E.: Adaptive responses of mammalian histidine-degrading enzymes. *J Biol Chem, 245:* 1204–1211, 1970.
86. Wergedal, J. E., Ku, Y., and Harper, A. E.: Influence of protein intake on the catabolism of ammonia and glycine in vivo. In Weber, G. (Ed.): *Advances in Enzyme Regulation.* Oxford, Pergamon Press, (New York, Macmillan), 1964, vol. II. pp. 289–299.
87. Peraino, C., Blake, R. L., and Pitot, H. C.: Studies on the induction and repression of enzymes in rat liver. III. Induction of ornithine transaminase and threonine dehydratase by oral intubation of free amino acids. *J Biol Chem, 240:* 3039–3043, 1965.
88. Haider, M., and Tarver, H.: Effect of diet on protein synthesis and nucleic acid levels in rat liver. *J Nutr, 99:* 433–445, 1969.
89. Henriques, O. B., Henriques, S. B., and Neuberger, A.: Quantitative aspects of glycine metabolism in the rabbit. *Biochem J, 60:* 409–424, 1955.
90. Black, A. L.: Modern techniques for studying the metabolism and utilization of nitrogenous compounds, especially amino acids. In *Isotope Studies on the Nitrogen Chain.* International Atomic Energy Agency, Vienna, 1968, pp. 287–309.
91. Leung, P. M. B., Rogers, Q. R., and Harper, A. E.: Effect of amino acid imbalance on plasma and tissue-free amino acids in the rat. *J Nutr, 96:* 303–318, 1968.
92. Peng, Y., Tews, J. K., and Harper, A. E.: Amino acid imbalance, protein intake and changes in rat brain and plasma amino acids. *Am J Physiol, 222:* 314–321, 1972.
93. Harper, A. E.: Effect of variations in protein intake on enzymes of amino acid metabolism. *Can J Biochem, 43:* 1589–1603, 1965.
94. Harper, A. E., Benevenga, N. J., and Wohlhueter, R. M.: Effects of disproportionate amounts of amino acids. *Physiol Rev, 50:* 428–558, 1970.
95. Peng, Y., and Harper, A. E.: Amino acid balance and food intake: Effect of amino acid infusions on plasma amino acids. *Am J Physiol, 217:* 1441–1445, 1969.
96. Beaton, J. R.: Methionine toxicity in rats exposed to cold. *Can J Physiol Pharm, 45:* 329–333, 1967.
97. Peng, Y., and Harper, A. E.: Amino acid balance and food intake: Effect of amino acid infusions on plasma and liver amino acids. *Am J Physiol, 217:* 1441–1445, 1969.
98. Leung, P. M. B., and Rogers, Q. R.: Importance of prepyriform cortex in food intake response of rats to amino acids. *Am J Physiol, 221:* 929–935, 1971.
99. Anderson, H. L., Benevenga, N. J., and Harper, A. E.: Effect of prior high protein intake on food intake, serine dehydratase activity and

plasma amino acids of rats fed amino acid imbalanced diets. *J Nutr, 97:* 463–474, 1969.

100. Anderson, H. L., Benevenga, N. J., and Harper, A. E.: Effect of cold exposure on the response of rats to a dietary amino acid imbalance. *J Nutr, 99:* 184–190, 1969.

101. Morris, M. L., Lee, S-C., and Harper, A. E.: Influence of differential induction of histidine catabolic enzymes on histidine degradation *in vivo. J Biol Chem, 247:* 5793–5804, 1973.

Chapter 6

NITROGEN BALANCE, INSULIN, AND AMINO ACID METABOLISM

T. T. Aoki, M. F. Brennan, W. A. Muller
and G. F. Cahill, Jr.

This presentation will be primarily concerned with interrelationships between body carbohydrate stores, muscle protein, insulin, and the overall nitrogen economy. In addition, the possible role of circulating blood cells of man in amino acid metabolism will be briefly discussed.

Free forms of carbohydrate (blood glucose) and protein (blood amino acids) are very minor sources of energy. Storage forms of these two fuels require an aqueous milieu and consequently contain relatively little energy per gram of wet tissue, i.e. 1 to 2 calories per gram.*

In man, the half-day supply of glycogen stored in muscle and liver is primarily reserved for interprandial glucose homeostasis and for emergency situations. In contrast, protein in man is only present in forms subserving a biological function such as oncotic pressure, biochemical catalysis, and structure. Thus protein is only indirectly available as a fuel and then only by proteolysis of muscle or other body proteins. Man must therefore depend on destruction of func-

Supported in part by USPHS Grants AM-05077, RF-31, AM-15191, HE 13872, The John A. Hartford Foundation, Inc., and the U.S. Army Research and Development Command DA-49-193-MD-2337.

* Lipid, the major energy source of man, will not be discussed, even though it is recognized that its high energy density (6 to 8 calories per gram) permits maneuverability, the key to survival in the animal kingdom.

tioning protein, principally muscle for glucogenic presursors when neither glucose nor protein are available from the immediate environment. This latter measure is necessitated by a limited supply of stored carbohydrate and the inability of animal systems to synthesize glucose from acetate derived from fatty acids, a capability found only in the plant kingdom.

Since limited stores of liver glycogen are rapidly exhausted in briefly fasted man, (1) they are incapable of maintaining glucose homeostasis for more than a few hours. Hepatic gluconeogenesis, primarily from amino acids, then becomes the mainstay of the blood glucose level and permits continued function of the major consumer of glucose in man, the brain. Rates of splanchnic glucose output in man vary greatly depending on nutritional status. Transition from fed to fasted states is associated with rapidly diminishing splanchnic glucose output and peripheral uptake. Current estimates suggest that splanchnic glucose production is approximately 180 grams per day, half being derived from amino acids. These estimates are in agreement with data of Pozefsky *et al.* (2) and others (3) who showed directly that splanchnic glucose production is primarily a function of amino acid release from muscle. Even with allowances for errors in calculation of muscle blood flow and mass, release of α-amino nitrogen from muscle approximates the quantity of glucogenic precursor needed to maintain glucose levels early in starvation. Remaining glucose is derived from glycerol released from adipose tissue, approximately 20 grams per day, and from residual liver glycogen. While some glucose, about 40 grams per day, is available from recycled lactate and pyruvate through the Cori Cycle, no net glucose is produced from this source. With more prolonged starvation, blood glucose and insulin levels decline, while increases occur in free fatty acid concentrations and hepatic ketoacid production. Most importantly, the brain begins to use β-hydroxybutyrate and acetoacetate as 60 to 80 percent of its fuel in lieu of glucose. Coincident with these changes in concentrations of levels of circulating substrates, urinary nitrogen excretion and hepatic gluconeogenic rates decrease.

Glutamine and alanine, both excellent glucogenic precursors, are the two amino acids released in largest amounts from muscle after either brief or prolonged fasting in man. (4) Glutamine is partially derived from proteolysis of muscle protein as well as from glutamate absorbed by muscle. Glutamate is then amidated by glutamine synthetase. (5) The amino group for this process comes either from deamination of branched-chain amino acids oxidized in muscle (6) or from aspartate by the purine nucleotide cycle of Tornheim and Lowenstein. (7) Alanine is released from muscle in part from proteolysis and indirectly by transamination from other amino acids. Alanine arises together with pyruvate either from glucose or Kreb Cycle intermediates or both because of small amounts of muscle phosphoenolpyruvate carboxykinase activity. (8) It is important to point out that if alanine were formed only from glucose, no net glucogenic carbons would be provided, and alanine could serve only to transport ammonium. Felig *et al.* (9) and Mallette *et al.* (10) have indeed described this pyruvate-glucose-alanine interrelationship as a cycle serving as an ammonium transport system.

Insulin is the principal and perhaps the only hormone acutely regulating both muscle proteolysis and amino acid availability. This concept is tenable because the first irreversible step in a sequence of reactions is usually the site of control. A feedback loop can thus be envisaged in which either circulating concentrations of glucose or amino acids or both regulate β-cell release of insulin. In turn, insulin suppresses muscle proteolysis which then inhibits both glutamine and alanine release from muscle. As blood glucose levels fall in briefly fasted man, insulin release and concentrations decline. Muscle then breaks down releasing amino acids into the circulation.

If small amounts of insulin are continuously infused into fasting subjects, plasma amino acid and blood glucose concentrations decline in concert with diminished urea nitrogen excretion. (11) These declines are evidence for decreased hepatic gluconeogenesis. Thus, in fasted man, hepatic gluconeogenesis is not limited by decreased enzymatic activity. Rather, more elegantly, hepatic gluconeogenesis is limited by decreased substrate, the glucogenic amino acids. It follows that elevated plasma alanine levels should then result in increased glucose levels. Felig *et al.* (12) have shown this to be true.

Dramatic and swift atrophy of muscle in a denervated or injured limb has been observed by many clinicians. Equally striking is selective hypertrophy of muscles exercised excessively while other muscles in the same hormone-fuel milieu do not. Goldberg (13) has found that when rat gastrocnemius tendon is divided, compensatory hypertrophy of adjacent ipsilateral soleus muscle occurs, even during starvation. Muscle work thus overrides insulin's control of muscle proteolysis and protein synthesis. These findings coincide neatly with two other clinical observations. Patients with insulin secreting tumors do not exhibit muscle hypertrophy. Also, well controlled diabetics have normal musculature. In sum, insulin integrates total muscle mass into glucogenic body needs, while individual muscle mass is determined by its cellularity and work load.

Cuatrecasas (14) has recently resolved the question as to whether insulin must enter the cell in order to initiate certain metabolic processes. Using agarose-bound insulin, he found that insulin need not enter adipose cells to initiate metabolic effects commonly attributed to it. Coupling of the insulin molecule to cell membrane is all that is required. Further, using affinity-chromatographic techniques, he characterized binding constants of insulin to fat cells and found that the concentration of insulin needed to bind half of the insulin receptors was identical both in fed and in fasting animals. (15) From these data, it is reasonable to propose that low insulin concentrations stimulate muscle proteolysis which, in turn, increases amino acid availability to other organs. On the other hand, high insulin levels stimulate protein synthesis. Finally, Livingston, Cuatrecasas, and Lockwood (16) have shown that even though fat cells of obese animals are insulin insensitive, they have a normal complement of receptor sites and demonstrate normal binding affinity. These data suggest that insulin resistance develops at some point between the receptor site and site of physiologic response. This observation is of great interest to those studying insulin resistance after trauma.

Following reaction of insulin with cell membrane, suppression of intracellular proteolysis occurs. Events leading to this suppression are unknown. However, it is known that intravenous administration of small amounts of insulin to fasting man results in decreased plasma amino acid levels. Decreased blood glucose concentrations and decreased urea nitrogen excretion follow without significantly altered blood ketoacid or free fatty acid levels. (11) The antiproteolytic effect is thus one of the most sensitive effects of insulin in man. This very sensitivity of muscle to insulin permits fasting man to conserve nitrogen while simultaneously mobilizing and using both free fatty acids and ketoacids, β-hydroxybutyrate and acetoacetate.

Discussion of nitrogen balance and amino acid balance in man has been centered on the muscle cell and plasma compartment. In 1969, Pozefsky *et al.* (2) infused insulin into the brachial artery of normal man for 26 minutes. They found a progressive decrease in the rate of release of both α-amino nitrogen and individual amino acids from forearm muscle. Since those amino acids lost by muscle to maintain glucose homeostasis must be replaced, a net accumulation of amino acids was expected. However, this net uptake of amino acids from plasma was not observed.

Recently, the possibility was investigated that higher insulin concentrations delivered for a longer time were needed to induce an accumulation of amino acids by muscle. Insulin was infused locally into forearm muscle of normal man for 90 minutes raising insulin concentrations within the forearm to approximately 200 micro units per ml (Normal concentrations are 10 to 20 micro units per ml). In addition, to assess the possible role of circulating blood cells in amino acid transport and exchange, whole blood, as well as plasma amino acid determinations were performed. Since a specific and highly reproducible assay for glutamate was available, special attention was directed to this amino acid. No changes were observed in *plasma* glutamate brachial artery—deep venous differences (A-DV) across forearm muscle during the intrabrachial artery insulin infusion. However, determination of *whole blood* glutamate brachial artery-deep venous differences revealed a significant uptake of glutamate during the insulin infusion. It was subsequently determined that before the insulin infusion, a net transfer of glutamate from arterial plasma into the blood cells occurred as blood traversed the muscle capillary bed. During the insulin infusion, however, both arterial blood cells and plasma lost glutamate to muscle. Thus, circulating blood cells, in contrast to previous theories, play an active role in both amino acid transport and metabolism. Studies now in progress have shown other amino acids to be affected by blood cell-plasma-tissue interchange. Whether abnormalities in this interrelationship occur after trauma is not known.

Preliminary studies at the Peter Bent Brigham Hospital of severely traumatized patients have shown the following. (1) Plasma amino acid levels are depressed.* (2) Arterio-deep venous plasma amino acid differences across forearm muscle are not very different from controls.* (3) Plasma immunoreactive glucagon concentrations are increased. (4) Finally, the T½ of alanine, i.e. the time for the specific activity of ^{14}C- (L) -alanine (U) given as an intravenous bolus to decrease by one half, a measure of hepatic gluconeogenesis, is very much shortened. (17) Further documentation of the hypercatabolic state of trauma stems from work showing increased glucose turnover and oxidation after injury. (18) This data is further evidence that amino acids released from muscle at an accelerated rate are converted into glucose by liver. Since reparative tissues use only glucose for substrate, it is reasonable to assume that trauma initiates muscle

---

* (Aoki, Brennan, Muller, Marliss, Cahill—Unpublished data).

hypercatabolism to provide fuel for repair. Nature did not plan on the availability of intravenously administered dextrose. In fact, if glucose is infused, adrenergic suppression of insulin release and peripheral insulin resistance, induced by undefined mechanisms, prevents use of excess exogenously administered glucose. Further, these mechanisms fail to suppress overproduction of endogenously synthesized glucose. As a consequence hyperosmolar coma may ensue, Hinton *et al.* (19) have shown that very large doses of insulin, 200 to 600 units per day, are capable of partially suppressing the hypercatabolic state of trauma. As evidence, they cite pronounced reduction of urinary urea nitrogen excretion in severely burned patients treated in this manner.

SUMMARY

Muscle provides glucogenic amino acids for hepatic gluconeogenesis in fasting man. Evidence is consistent with insulin being the controlling signal operating by a negative feedback mechanism. Exercise of a given muscle may alter the insulin-muscle metabolism relationship. Circulating blood cells may play an important role in amino acid transport between tissues. Trauma accentuates muscle proteolysis, but very high concentrations of insulin are capable of suppressing this hypercatabolic state.

REFERENCES

1. Hultman, E., and Nilsson, L. H.: Liver glycogen in man. Effects of different diets and muscular exercise. In Pernow, B., and Saltin, B., (Eds.): *Muscle Metabolism in Exercise,* New York, Plenum, 1971, pp. 143–151.
2. Pozefsky, T., Felig, P., Tobin, J. D., Soeldner, J. S., and Cahill, G. F., Jr.: Amino acid balance across tissues of the forearm in postabsorptive man. Effects of insulin at two dose levels. *J. Clin Invest, 48:* 2273, 1969.
3. London, D. R., Foley, T. H., and Webb, C. G.: Evidence for the release of individual amino acids from resting human forearm. *Nature, 208:* 588–9, 1965.
4. Marliss, E. B., Aoki, T. T., Pozefsky, T., Most, A. J., and Cahill, G. F., Jr.: Muscle and splanchnic glutamine and glutamate metabolism in postabsorptive and starved man. *J Clin Invest, 50:* 814, 1971.

5. Iqbal, K., and Ottaway, J. H.: Glutamine synthetase in muscle and kidney. *Biochem J, 119:* 145, 1970.
6. Manchester, K.: Oxidation of amino acids by isolated rat diaphragm and the influence of insulin. *Biochim Biophys Acta, 100:* 295–298, 1965.
7. Tornheim, K., and Lowenstein, J. M.: The purine nucleotid cycle. The production of ammonia from aspartate by extracts of rat skeletal muscle. *J Biol Chem, 247:* 162, 1972.
8. Opie, L. H., and Newsholme, E. A.: The activities of fructose 1,6-diphosphatase, phosphofructokinase, and phosphoenolpyruvate carboxykinase in white muscle and red muscle. *Biochem J, 103:* 391–99, 1967.
9. Felig, P., Pozefsky, T., Marliss, E. B., and Cahill, G. F., Jr.: Alanine-key role in gluconeogenesis. *Science, 167:* 1003, 1970.
10. Mallette, L. E., Exton, J. H., and Park, C. R.: Control of gluconeogenesis from amino acids in the perfused rat liver. *J Biol Chem, 244:* 5713, 1969.
11. Cahill, G. F., Jr.: Physiology of insulin in man. *Diabetes, 20:* 785–799, 1971.
12. Felig, P., Owen, O. E., Wahren, J., and Cahill, G. F., Jr.: Amino acid metabolism during prolonged starvation. *J Clin Invest, 48:* 584–94, 1969.
13. Goldberg, A. L.: Relationship between hormones and muscular work in determining muscle size. In Alpert, N. (Ed.): Cardiac Hypertrophy. New York, Academic Press, 1971, p. 39.
14. Cuatrecasas, P.: Interaction of insulin with the cell membrane, the primary action of insulin. *Proc Nat Acad Sci (USA), 63:* 450–7, 1969.
15. Cuatrecasas, P.: Insulin-receptor interactions in adipose tissue cells: Direct measurement and properties. *Proc Nat Acad Sci (USA), 68:* 1264–8, 1971.
16. Livingston, J. N., Cuatrecasas, P., and Lockwood, D. H.: Insulin insensitivity of large fat cells. Submitted to *Science,* 1972.
17. Marliss, E. B., Aoki, T. T., Morgan, A. P., O'Connell, R. C., and Cahill, G. F., Jr.: Alanine in catabolic states. *Clin Res, 18:* 539, 1970.
18. Long, C. L., Spencer, J. L., Kinney, J. M., and Geiger, J. W.: Carbohydrate metabolism in man: Effect of elective operations and major injury. *J Appl Physiol, 31:* 110–16, 1971.
19. Hinton, P., Allison, S. D., Littlejohn, S., and Lloyd, J.: Insulin and glucose to reduce catabolic response to injury in burned patients. *Lancet, 1:* 767–9, 1971.

CHAPTER 7

PROGRESS IN PARENTERAL PROTEIN NUTRITION

ROBERT L. RUBERG
EZRA STEIGER
CHARLES VAN BUREN
STANLEY J. DUDRICK

PROGRESS IN UNDERSTANDING protein metabolism and particularly newer techniques as hyperalimentation have made intravenous protein feeding more effective and relatively safe.

In this chapter, three areas of parenteral protein feeding will be reviewed:

1. Events which led to development of current intravenous alimentatation techniques.
2. Basic principles involved.
3. New directions of research.

HISTORICAL REVIEW

The first really successful protein hydrolysate solution for intravenous infusion into patients was prepared from casein by Elman and Wiener in the late 1930's. (1) Prior to that time, only whole protein as blood or its products had been used. Elman and Wiener achieved limited success with positive nitrogen balance, but adverse reactions to these early solutions prevented routine use. Their efforts, however,

---

Supported in part by U.S. Army Contract #DADA17-72-C2026 and USPHS Grant GM-01540-06.

did show that intravenous protein hydrolysate feeding was both feasible and beneficial for patients unable to take food orally. Later, in the 1940's, Werner, (2) using solutions of highly-purified amino acids also had modest success in treating clinical disorders complicated by malnutrition.

Available protein hydrolysates theoretically might have promoted protein anabolism. One limiting factor was inability to infuse enough calories to permit the hydrolysates to be used for new protein synthesis. Seriously ill patients required many more calories than could be delivered reasonably with 5% (isotonic) dextrose solution. In the early 1960's, Rhoads and co-workers (3) used slighly concentrated dextrose solutions, 10 to 15%, fat emulsions, and protein hydrolysates in combination with intravenous diuretics to achieve positive nitrogen balance in some surgical patients. Because of large fluid volume, usually 7 liters per day, this technique was impractical for many patients.

Two problems then confronted investigators. (1) Excessive fluid volume was required for adequate total parenteral nutrition. (2) Infusion of more concentrated solutions into peripheral vessels caused phlebitis and venous thrombosis.

These problems have been solved in part with the evolution of intravenous hyperalimentation. This technique may be described by stating principles upon which it is based: (1) The parenteral diet is carefully formulated, purified, and highly concentrated. (2) Average normal fluid requirements are infused with high concentrations of both calories and protein. (3) A calorie to nitrogen ratio of 150:1, i.e. 150 calories per gram of nitrogen, or higher, is used in an effort to minimize degradation of amino acids for energy. (4) The hyperosmolar solution is immediately diluted, avoiding phlebitis and thrombosis by infusing it into a large-bore central vein. By infusing at a constant rate over 24 hours, nutrients are better used and their renal excretion is lessened. In contrast to normal oral feeding, continuous rather than intermittent intravenous feeding is more effective.

This method for complete intravenous nutrition was first used to achieve normal growth and development in beagle puppies. (4) Later, a similar, equally effective, system was used in infants. (5) In adults, hypertonic parenteral feeding through a subclavian vein catheter was often associated with positive nitrogen balance, wound healing, and weight gain. (6,7) Refinements in solution preparation, delivery systems and patient monitoring have made intravenous feeding safer and more practical. Since these original demonstrations, more than 1,000 patients at our institution and very many others elsewhere have been maintained with this technique.

In most disease states, caloric and nitrogen requirements are greater than normal. Because greater than normal amounts are infused to meet these needs, the term intravenous hyperalimentation is used. Each liter of standard solution for this method contains 200 to 250 grams of glucose and 40 to 50 grams of hydrolyzed protein, usually fibrin or casein. Two to 3.5 liters of this solution daily, the average amount infused if renal function is normal, contain 2,000 to 3,500 calories and 75 to 130 grams of protein hydrolysate. These requirements are exceeded in hypermetabolic states as after severe burns or trauma, and larger volumes are administered if tolerated. Indications for use of this technique as well as listing of conditions where it may not be used are to be found in many review articles on this topic. (16,17) *

Many investigators are studying ways to improve the original regimen. Discussions of our efforts will comprise the remainder of this paper.

NEW SOLUTIONS

To improve parenteral nutrition solutions, better calorie sources must be found as well as better balanced and more refined protein and amino acid mixtures.

---

* Editor's note: In brief, principal reasons for not using or discontinuing the technique are incontrollable sepsis at site of infusion catheter, uncontrolled electrolye or osmolar abnormalities, or for patients with terminal hopeless illness.

Glucose, the major caloric source, is sometimes inadequate. Other sugars as the pentose, xylitol (18) and the hexose, sorbitol, (19) have been studied by others for their nutritional value alone or with glucose.

Fat emulsions as, for example, those prepared from soybean oil, have been tried as a principal calorie source. Difficulties from adverse reactions are under study, but these fat emulsions have not been approved for general clinical use in this country.

Serum essential fatty acid deficiencies often occur with prolonged intravenous hyperalimentation. In a large series of patients we have observed low concentrations of linoleic acid, a 17 carbon unsaturated indispensible fatty acid, and high values for Δ 5, 8, 11-eicosatrienoic acid, a 20 carbon chain. These abnormalities stress importance of finding satisfactory intravenous fat emulsions and suggest need for study of their routine administration.

Principal efforts to improve intravenous protein nutrition have been to find solutions of pure amino acids to replace protein hydrolysates. Recently, solutions of pure crystalline amino acids have become available at reasonable cost. Such solutions have several theoretic advantages over standard fibrin or casein hydrolysates. First, exact amino acid composition is known. Solutions may be tailored to fit individual needs in contrast to the fixed amino acid composition of protein hydrolysates. Further, study may be pursued for "ideal" or optimal combinations of essential and non-essential amino acids.

A second advantage of synthetic amino acid mixtures is purity. Present casein or fibrin hydrolysates contain significant amounts of di-, tri-, and even larger peptides from incomplete hydrolysis. Febrile and allergic responses which occur occasionally are believed due to these peptides. Finally, crystalline amino acids are available in larger quantities than the relatively limited supplies of casein and fibrin.

Pure crystalline amino acid solutions have been used in other countries. Several are being readied for possible commercial release in this country. Recently, we have

TABLE 7-I

	*Fibrin Hydrolysate** gm/L	*Crystalline Amino Acids*** gm/L
l-Lysine	4.00	4.00
l-Tryptophan	0.50	0.50
l-Phenylalanine	1.00	1.00
l-Methionine	1.00	1.00
l-Leucine	6.36	6.36
l-Isoleucine	2.18	2.18
l-Valine	1.63	1.63
l-Threonine	2.32	2.32
l-Arginine	2.90	2.90
l-Histidine	1.16	1.16
l-Glycine	2.08	2.08
l-Cysteine	0.30	....
l-Glutamic Acid, l-Tyrosine, l-Aspartic Acid	5.57	....
Peptides	19.00	....
	50.00	25.13

* Aminosol®
** Neoaminosol®

tested safety and efficacy of three amino acid solutions clinically. All were reasonably successful in promoting positive nitrogen balance. Their infusion was associated with wound healing and weight gain. In one series of experiments, (8) a formulation* based on the amino acid profile of standard fibrin hydrolysate** was infused for periods of 5 to 41 days in 20 patients (Table 7-I). Each liter contained 25 grams of crystalline amino acids and 200 grams of glucose with vitamins and minerals. This liter potentially provided 900 calories. Average duration of total parenteral feeding was 16 days. The mean infusion was 3.1 liters per day or about 75 grams of amino acids, 600 grams of glucose and 2700 calories. Patients gained an average of 6 pounds during the intravenous feeding period. Nineteen of 20 patients had a positive nitrogen balance. Our impression was that this regimen was usually consistent with accelerated wound healing and decreased postoperative morbidity. Use of the two other amino acid-dextrose solutions gave similar results.

* NeoAminosol®, Abbott
** Aminosol®, Abbott

Alterations in calcium-phosphorus metabolism as well as glycolysis, 2-3 diphosphoglycerate levels and the oxygen-hemoglobin curve of red cells were also studied. (9,10) As a result of these studies, difficulties of hyperalimentation associated with hypophosphatemia, as well as red cell membranes, particularly in regard to glycolysis and 2-3 diphosphogycerate are now better controlled.

Until the present time, study of amino acid infusion has centered on nitrogen balance measurements. More precise information for need of each amino acid in terms of concentration may be given by studying balance of each individual amino acid. The expression of concentration of each individual amino acid in urine or plasma is also called an amino acid profile. Data now being gathered from comparison of these two profiles may be useful for improving amino acid composition of intravenous feeding solutions.

NEW INDICATIONS FOR PARENTERAL FEEDING

Recently, administering intravenous hypertonic dextrose and essential L-amino acids was shown to benefit renal failure patients. (11) Use of this diet is based on data of Rose and Dekker (12) showing that when essential amino acids are provided, urea is used for non-essential amino acid synthesis in growing rats. Later, Giordano (13) and Giovannetti (14) in agreement with these observations showed that chronically uremic patients also synthesized protein when fed diets containing minimal requirements of essential L-amino acids and adequate calories. The Giordano-Giovannetti diet is now generally accepted for feeding uremic patients not having ready access to dialysis or kidney transplantation.

We have used a similar, highly purified intravenous diet in more than 40 patients. They were both surgical and non-surgical with impaired renal function and could not ingest an adequate diet orally. This diet contained eight essential L-amino acids formulated according to Rose (12) (Table 7-II). Most energy from this diet for protein synthesis is supplied by 50 to 75 percent dextrose solution. By using

TABLE 7-II

Renal Failure Solution (per 100 ml)	
l-Isoleucine	0.70 gm
l-Leucine	1.10 gm
l-Lysine	0.80 gm
l-Methionine	1.10 gm
l-Phenylalanine	1.10 gm
l-Threonine	0.05 gm
l-Tryptophan	0.25 gm
l-Valine	0.80 gm
	6.35 gm

these very concentrated solutions enough calories and amino acids can be provided even within fluid restriction of severely oliguric or anuric patients. In general, blood urea nitrogen (BUN) concentrations declined 40 to 50 percent when initial levels were of the order of 100 mg percent. A relatively low BUN was maintained if therapy had started immediately after dialysis. Equilibrium of potassium, nitrogen and water balance with stable or decreasing BUN suggests protein synthesis from endogenous urea nitrogen. (13,14) In these circumstances, weight remains constant or increases during the intravenous feeding period.

Even though clinical results were favorable, experimental control was not possible. One might argue that spontaneous reversal of renal failure occurred independently of treatment but coincidentally with intravenous therapy.

An experiment was designed to control this variable. (20) Beagle dogs were protein-depleted for two weeks to induce protein catabolism. A bilateral nephrectomy was then performed on each under ether anesthesia. At surgery, a plastic catheter was placed in an external jugular vein through a small incision (cutdown site) for central intravenous feeding into the superior vena cava. Dogs were placed in metabolic cages and attached to a swivel infusion apparatus which allowed unrestrained movement. Peristaltic pumps delivered intravenous diets at constant rates continuously.

Anephric dogs were divided into four experimental groups, each receiving the same fluid volume per day. The first group of eight animals was fed a diet comparable to the intravenous human renal failure regimen, that is, a mixture of 56 percent glucose and 0.525 gm/kg of essential L-amino acids. The second group of seven received an isocaloric dextrose diet totally devoid of amino acids. The third group of six was given an isovolemic (equal volume) diet consisting of 5 percent dextrose. The latter group might be considered representative of conventional therapy in surgical patients with renal failure. They received simple volume for volume fluid replacement with dextrose solutions lacking amino acids. A fourth group of six beagles offered water and standard kennel rations ad lib, served as controls.

This study illustrated differences in survival times of anephric animals. Group I animals lived longest, survival averaging 10 days. Group II animals averaged 8 days, while group III and control animals lived only 4.5 days. Differences between Groups I and II were not statistically significant. Intravenous infusion of concentrated dextrose solution with or without amino acids, nevertheless, was associated with a lower BUN, presumably through use of urea nitrogen for synthesis of non-essential amino acids. Even though data do not show increased survival from amino acid infusion, results do indicate better use of nitrogen with the amino acids.

Serum creatinine increased progressively to very high concentrations in all animals. Conversely, concentrations of blood urea nitrogen differed significantly. In Groups II and III, animals receiving dextrose without amino acids, BUN concentrations increased rapidly to 140 mg percent or more. In animals receiving essential amino acids with dextrose, however, BUN concentrations increased early but then remained relatively constant in the 40 to 60 mg percent range. These data are consistent with clinical observations that in patients unable to eat, the intravenous Girodano-Giovannetti diet can be very useful in treating renal failure.

Hepatic failure patients requiring intravenous feeding theoretically also might benefit from a synthetic amino acids solution. Standard hydrolysates of casein or fibrin usually contain 7 to 8 percent nitrogen as ammonium. They usually are not given to these patients because high ammonium concentrations are associated with hepatic encephalopathy. In contrast, synthetic crystalline essential and non-essential amino acid solutions contain no significant amounts of ammonium. For this reason, the purified essential L-amino acid diet in combination with 20 to 50 percent glucose was administered to a number of patients with liver disorders. Early success has been modest, but it is still relatively early in this study. Intravenous administration of one unit of crystalline amino acids (52.5 gm) was associated with normal blood ammonium concentrations, two units was associated with moderate elevation and three units with abnormally high blood ammonium levels. This suggests a defect in amino acid metabolism in these patients. (21)

NEW EXPERIMENTAL MODELS

Previous studies in parenteral feeding have been principally in dogs and humans. Limitations of human investigation are numerous. Likewise, dog experiments require much effort, expense, space, and time. Recently, a simple, inexpensive and very workable model for prolonged intravenous feeding in rats has been devised. The model closely duplicates parenteral hyperalimentation techniques learned earlier from human and dog work.

The procedure consists of threading a tiny plastic catheter through a small incision (cutdown) into the rat's jugular vein then into the superior vena cava. The proximal end of the catheter is then tunneled under the skin to exit between the scapulae, and is supported by a harness which permits attachment to a continuous infusion apparatus. Outside metabolic cages, tubing is connected to a swivel apparatus permitting the animal virtually unrestricted motion. A pump delivers the nutritional solution at a constant rate through the swivel apparatus.

This system has several advantages. First, much background information from oral nutrition studies in the rat is available to aid in interpretation of data. Second, space and time required are considerably less than for dogs. Third, control variations in diet and experimental design easily can be made. Rats have been maintained entirely by intravenous feeding for five weeks.

Early experiments measured postoperative nutrition, i.e. hyperalimentation on colonic wound healing, serum albumin, and body weight. (15,22) Sprague-Dawley rats were divided into three groups of 10 each and protein depleted for 6 weeks. Under anesthesia, each received a standard colon anastomosis and was catheterized for intravenous feeding. Group I rats were fed 30 percent dextrose and 5 percent amino acids. Group II received an isocaloric diet of 30 percent dextrose without amino acids. Group III rats were fed an isovolemic regimen of 5 percent dextrose alone. Vitamins and minerals were given to all animals. After 7 days, serum albumin, body weight and colonic bursting strength, using a standardized method, were measured.

Group I animals, fed a diet analogous to the usual adult human hyperalimentation regimen, gained an average of 29 grams of total body weight, while the other two groups remained the same or lost weight. Colon bursting strength was interpreted to be greatest in Group I. Total circulating serum albumin had returned to near normal in Group I animals while remaining low in Groups II and III.

Analysis of data from Groups II and III suggested that intravenous feeding with 5 percent glucose alone was more beneficial than 30 percent glucose without amino acids. Although Group III animals lost weight, serum albumin concentrations were greater and animals were not edematous as were Group II. In addition, microscopic studies showed relatively normal liver architecture in Group I and III, but fatty infiltration in Group II. A more complete resume of these and later experiments is given in Table 7-III.

These results with rats are consistent with clinical observations that hyperalimentation techniques including use of amino acids can benefit nutritionally depleted patients immediately following surgery. Perhaps more importantly, these experiments illustrate potential of this model for future experimentation.

TABLE 7-111

Effects of Postoperative Nutrition on Serum Proteins, Body Weight and Colonic Wound Healing

	IV Hyper-alimen-tation (1)	*Oral Hyper-alimen-tation (4)*	*IV Hyper-tonic (30%) Dextrose (2)*	*Oral Hyper-tonic (30%) Dextrose (5)*	*IV 5% Dextrose (3)*	*Oral 5% Dextrose (6)*
No. of animals	10	10	10	10	10	10
Weight (gm)	+ 17	+ 17	− 11	− 13	− 48	− 29
Total caloric intake (cal/day)	67 ± 14	53 ± 19	47 ± 12	35 ± 15	9 ± 1	22 ± 8
Total serum proteins (gm/100 ml)	5.9 ± 0.4	6.0 ± 0.2	4.8 ± 0.5	5.2 ± 0.2	5.5 ± 0.8	5.5 ± 0.5
Serum albumin (gm/100 ml)	3.1 ± 0.4	3.2 ± 0.2	1.9 ± 0.2	2.6 ± 0.2	2.6 ± 0.7	2.5 ± 0.3
Total circulating serum albumin (gm)	0.24 ± .05	0.24 ± .02	0.13 ± .03	0.17 ± .04	0.17 ± .04	0.18 ± .04
Bursting strength #1 (mm Hg)	188 ± 18	195 ± 8	161 ± 10	159 ± 25	176 ± 10	161 ± 13
Bursting strength #2 (mm Hg)	218 ± 36	199 ± 19	216 ± 8	200 ± 22	185 ± 25	182 ± 25

From Daly, J. M., Steiger, E., and Dudrick, S. J.: Postoperative nutrition and colonic wound healing, serum protein metabolism and body weight. *Surg Forum, 23:* 38–40, 1972. By permission.

SUMMARY

Much progress has been made in the growing field of intravenous protein nutrition. The technique of parenteral hyperalimentation is safe and effective for nutritional therapy in patients with impaired gastrointestinal function. Refinements in the basic solution and development of specialized new amino acid solutions have extended indications as well as improved efficiency and efficacy. New experimental models, particularly with the rat, will help to point out metabolic events and to further define indications for intravenous protein therapy.

REFERENCES

1. Elman, R., and Weiner, D. O.: Intravenous alimentation: With special reference to protein (amino acid) metabolism. *JAMA, 112:* 796, 1939.
2. Werner, S. C.: The use of a mixture of pure amino acids in surgical nutrition. I. Certain pharmacologic considerations. *Ann Surg, 126:* 169, 1947.
3. Rhoads, J. E., Rawnsley, H. M., Vars, H. M., Crichlow, R. W., Nelson, H. M., Spagna, P., Dudrick, S. J., and Rhoads, J. E., Jr.: The use of diuretics as an adjunct in parenteral hyperalimentation for surgical patients with prolonged disability of the gastrointestinal tract. *Bull Inter Soc Surg, 24:* 59, 1965.
4. Dudrick, S. J., Rhoads, J. E., and Vars, H. M.: Growth of puppies receiving all nutritional requirements by vein. In *Fortschritte der Parenteralen Ernahrung.* Locham bei Munchen, West Germany, Pallas Verlag, 1967, vol. II, pp. 16–18.
5. Wilmore, D. W., and Dudrick, S. J.: Growth and development of an infant receiving all nutrients exclusively by vein. *JAMA, 203:* 860, 1968.
6. Dudrick, S. J., Wilmore, D. W., Vars, H. M., and Rhoads, J. E.: Long term total parenteral nutrition with growth, development and positive nitrogen balance. *Surg, 64:* 134, 1968.
7. Dudrick, S. J., Wilmore, D. W., Vars, H. M., and Rhoads, J. E.: Can intravenous feeding as the sole means of nutrition support growth in the child and restore weight loss in an adult? An affirmative answer. *Ann Surg, 169:* 974–984, 1969.
8. Ruberg, R. L., Dudrick, S. J., Long, J. M., Allen, T. R., Steiger, E., and Rhoads, J. E.: Pre- and post-operative nutrition using crystalline amino acids as the sole source of nitrogen. *Fed Proc, 30:* 300, 1970.

9. Travis, S. F., Sugerman, H. J., Ruberg, R. L., Dudrick, S. J., Delivoria-Papadopoulos, M., Miller, L. D., and Oski, F. A.: Alterations of red cell glycolytic intermediates and oxygen transport as a consequence of hypophosphatemia in patients receiving intravenous hyperalimentation. *New Eng J Med, 285:* 763, 1971.
10. Ruberg, R. L., Allen, T. R., Goodman, M. J., Long, J. M., and Dudrick, S. J.: Hypophosphatemia with hypophosphaturia in hyperalimentation. *Surg Forum, 22:* 87, 1971.
11. Dudrick, S. J., Steiger, E., and Long, J. M.: Renal failure in surgical patients. Treatment with intravenous essential amino acids and hypertonic glucose. *Surg, 68:* 180, 1970.
12. Rose, W. C., and Dekker, E. E.: Urea as a source of nitrogen for the biosynthesis of amino acids. *J Biol Chem, 223:* 107, 1956.
13. Giordano, C.: Use of exogenous and endogenous urea for protein synthesis in normal and uremic subjects. *J Lab Clin Med, 62:* 321, 1963.
14. Giovannetti, S., and Maggiore, Q.: A low nitrogen diet with protein of high biological value for severe chronic uremia. *Lancet, 1:* 1000, 1964.
15. Steiger, E., Allen, T. R., Daly, J. M., Vars, H. M., and Dudrick, S. J.: Beneficial effects of immediate postoperative total parenteral nutrition. *Surg Forum, 22:* 89, 1971.
16. Dudrick, S. J., and Rhoads, J. E.: New horizons for intravenous feeding. *JAMA, 215:* 939–49, (Feb. 8) 1971.
17. Shils, M. E.: Guidelines for total parenteral nutrition. *JAMA, 220:* 1721–1729, (June 26) 1972.
18. Chamberlain, J., et al.: D-xylose absorption before and after portasystemic venous anastomosis. *Lancet, 1:* 1030–1032, (May 24) 1969.
19. Bye, P. A.: Utilization and metabolism of intravenous sorbitol. *Brit J Surg, 56:* 653–656, (Sept.) 1969.
20. Van Buren, C. T., Dudrick, S. J., Dworkin, L., Baumbauer, E., and Langer, J. M.: Effects of intravenous essential L-amino acids and hypertonic dextrose on anephric beagles. *Surg Forum, 23:* 83–84, 1972.
21. Dudrick, S. J., McFayden, B. V., Van Buren, C. T., Ruberg, R. L., and Maynard, A. T.: Parenteral hyperalimentation: Metabolic problems and solutions. *Ann Surg, 176:* 259–264, (Sept.) 1972.
22. Daly, J. M., Steiger, E., and Dudrick, S. J.: Postoperative nutrition and colonic wound healing, serum protein metabolism and body weight. *Surg. Forum, 23:* 38–40, 1972.

CHAPTER 8

SUBSTRATE PROFILE IN PROTEIN WASTING STATES

George L. Blackburn and Jean-Pierre Flatt

Significant protein losses during disease were first shown by nitrogen balance studies at the turn of the century. For example, increased nitrogen losses were noted by Schaeffer and Coleman (1) in typhoid fever, by Coleman and Dubois (2) in erysipelas, and by Dubois in both malaria and arthritis. In 1932, Cuthbertson (3) reported that fracture of a femur led to increased nitrogen losses both in man and experimental animals. Thirteen years later, Munro showed in similar experimental animals that by feeding a protein-free diet, less nitrogen was lost. (4) During World War II and the Korean War many cases of protein losing states were documented renewing interest and emphasizing the importance of nutrition. (5) Mechanisms leading to extensive protein loss during infection and following injury are unknown, but a number of conflicting hypotheses have been advanced.

The purpose of this presentation is to document significant changes occurring in those blood substrate levels used for energy production. Studies of the complete substrate profile* including carbohydrate, fat, and amino acid metabolites,

Supported in part by U.S. Public Health Service Grants AM-14161, AM-08681, AM-5618.

* Substrate profile refers to peripheral venous blood glucose (glu), lactate (lac), pyruvate (pyr), free fatty acids (FFA), β-hydroxybutyrate (HB), acetoacetate (AcAc). The latter two collectively may be referred to as ketone bodies (KB). Key hormones involved in utilization of substrates are insulin (ins), epinephrine (epi), glucagon and cortisol.

have led us to propose a new approach to reduce protein losses of patients during periods of semistarvation.

This approach is based first on recognition of importance of adequate fat mobilization, including starvation ketosis as prerequisites for efficient protein sparing. It emphasizes new data showing that insulin, whose anabolic effect on protein metabolism is desirable, also exerts an unfavorable anabolic effect on lipid metabolism when caloric intake is insufficient.

PROTEIN METABOLISM IN TRAUMA AND SEPSIS

The total cellular protein of the body, excluding connective tissue and bone matrix, is about 8 kilograms. The greatest proportion is muscle protein. Other body proteins include a soluble protein pool, including plasma proteins, the extra cellular connective tissue protein, and skeletal protein.

Nitrogen balance refers to intake and excretion of nitrogen. Under normal conditions nitrogen excretion will equal intake. For this to occur the diet must supply sufficient calories and usually at least 0.5 grams of protein per kilogram of body weight per day. Most nitrogen is excreted in urine. The remainder is lost as fecal nitrogen, sweat, and cutaneous tissues. Urea is the major form of nitrogen excretion, with ammonia, uric acid, creatinine, amino acids, and some peptides making up the rest. No storage site for nitrogen compounds is known in the body. A positive nitrogen balance indicates overall predominance of synthesis over degradation of protein. The reverse is true when nitrogen balance is negative.

Lean tissue mass may gain or lose large amounts of protein without much alteration in concentration of total plasma proteins or of plasma albumin. Furthermore, protein can be synthesized in one site and broken down in another simultaneously. After trauma or hemorrhage, for example, wound proteins, hemoglobin, and albumin synthesis are extensive, whereas muscle tissue is breaking down.

Caloric needs are measured by a subject's oxygen consumption together with his carbon dioxide and nitrogen output. Multiple measurements have shown caloric needs to be

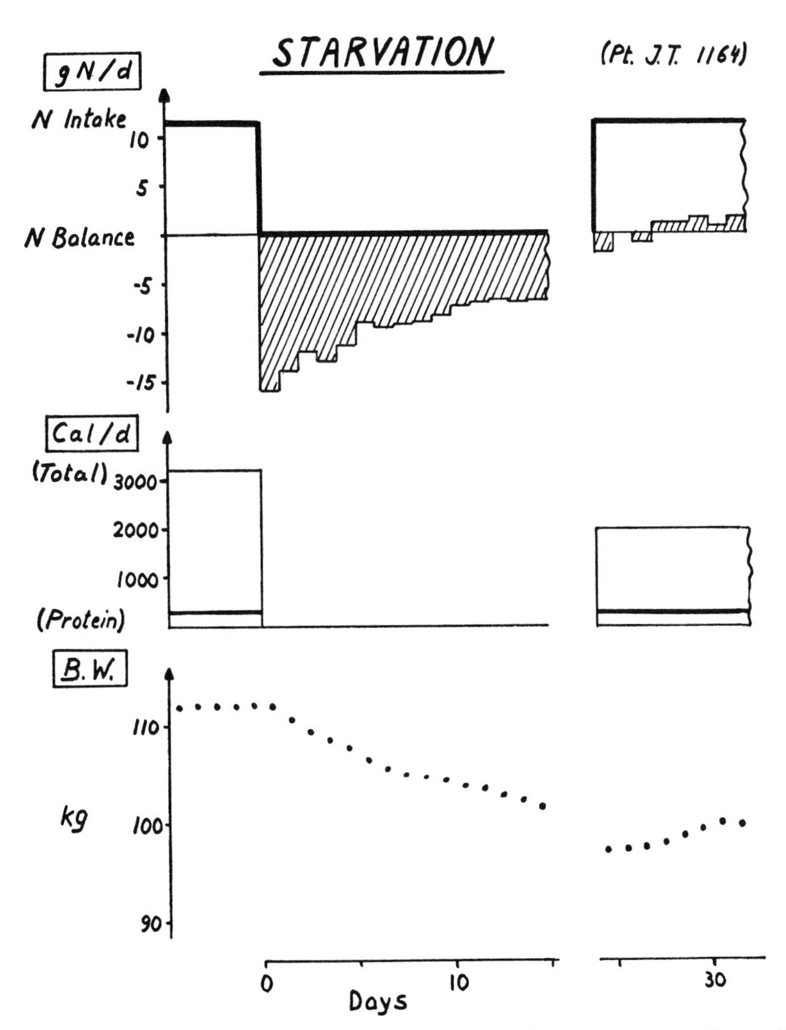

Figure 8-1. Decrease in protein catabolism during total starvation. An obese (112 kg body weight) male, 46 years old, underwent a period of total starvation for weight reduction. The figure shows daily protein nitrogen and caloric intake as well as the nitrogen balance (N intake minus urinary N loss). Changes in body weight are also shown.

20 to 40 percent greater than normal in patients with sepsis and trauma. (6) One exception is the burned patient whose requirements are much higher due to extensive evaporative fluid loss from burned surfaces. Very ill patients often cannot meet this requirement by oral feedings. An example is disruption of function of the gastrointestinal tract. In these instances the patient will be forced to mobilize endogenous energy stores such as glycogen, fat, and tissue protein. The latter is the most valuable to the organism. Determination of nitrogen balance will indicate how efficiently his metabolism is able to adapt to minimize protein losses. Kinney (6) has determined that mean total nitrogen excretion in preoperative, early operative, and septic groups of patients were 5.9, 5.2, and 7.1 g (nitrogen/square meter/day) respectively.

Figure 8-1 shows nitrogen balance in a healthy, obese patient undergoing a period of total starvation after a control period of a constant protein, isocaloric diet. Nitrogen loss decreases appreciably during each successive day of fast. A number of studies show that after 5 weeks of starvation only 4 to 5 grams of nitrogen are lost per day. (7) It is of interest that upon refeeding, positive nitrogen balance is achieved even when caloric intake remains below caloric requirement.

Much present understanding of metabolic behavior during starvation in man has been contributed by Cahill and his collaborators. In Figure 8-2, we use data which they report on the gradual decrease in nitrogen excretion. (21) When daily nitrogen excretion is plotted semilogarithmically, one finds the points falling in a straight line indicative of an exponential decline. It is not clear whether the half-life of 19.4 days and initial protein mass of 2.2 kilograms describing this exponential decline have a physiological meaning. This data, nevertheless, suggests that rate of protein loss is related, among other factors, to the mass of available body protein.

In contrast, during disease or trauma with negative caloric balance, the body is much less efficient in preserving its protein (6,8) as illustrated in Figure 8-3. This patient was

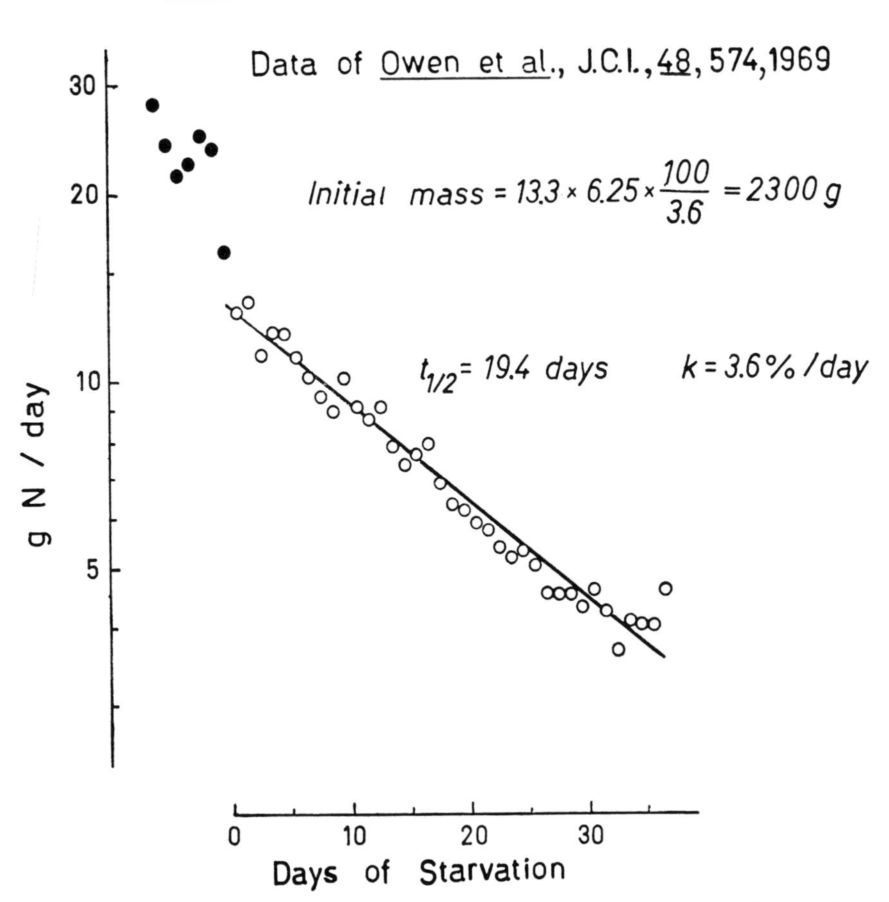

Figure 8-2. Urinary nitrogen excretion during starvation in man. The graph shows a semilogrithmic plot of urinary nitrogen output per day as a function of days of total starvation. The data are those reported by Owen *et al.*[21] for a 49-year-old male whose weight changed from 132 kg to 110 kg in 38 days.

a 65-year-old male suffering from perforated duodenal ulcer with peritonitis, complicated by advanced emphysema. Nitrogen losses increased over several days instead of decreasing as in patients of Figures 8-1 and 8-2. Even though this patient was receiving 400 calories daily in dextrose infusions, protein wasting was considerable until caloric intake in-

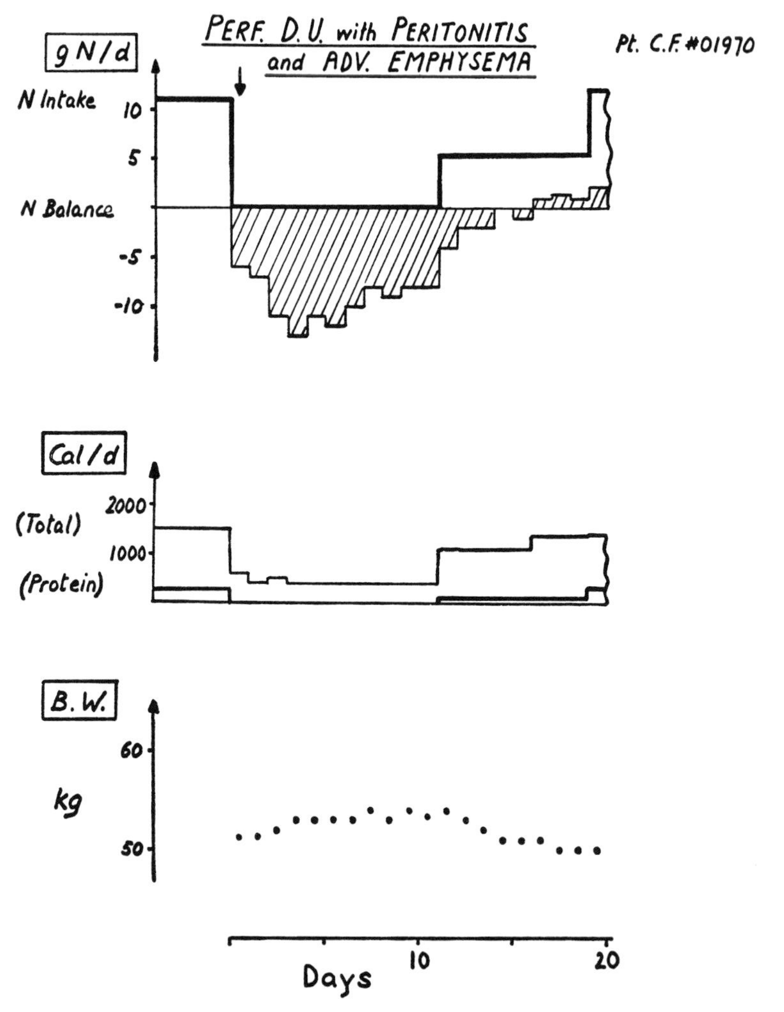

Figure 8-3. Protein wasting during trauma and sepsis. On the basis of his outpatient record this 65-year-old male was presumably in nitrogen balance when he suffered perforation of a duodenal ulcer, complicated by peritonitis. He also suffered from advanced emphysema. Considering his initial body weight of 51 kg, he lost considerable protein in spite of glucose infusions of 400 calories per day.

creased and resolution of peritonitis occurred. Protein wasting was particularly striking in view of his emaciation, his body weight being only 51 kilograms.

Indeed since Cuthbertson originally observed (3) increased urinary nitrogen excretion after long bone fractures, protein catabolism after trauma has been studied extensively. (6,8) It is widely held that increased nitrogen excretion after injury reflects mobilization of amino acids to meet increased demands for metabolic fuels. Stephens and Randall (9) and later, Border, (10) concluded that protein and fat became primary energy sources following injury. Moore (8) using serial measurements of body composition concluded that tissue loss during surgical convalescence is approximately half fat and half lean body mass by weight.

Protein breakdown is believed by some to be due in part to the requirement for glucose. (6) This need must be met through gluconeogenesis. Infusion of glucose would then be expected to reduce gluconeogenesis and nitrogen loss. Neither occurs with sepsis and trauma as shown in Figures 8-2 and 8-3 and documented by others. (6) Our measurement of caloric expenditure in patients with varying degrees of sepsis and trauma show that even with normal oral or parenteral caloric intake, they remain in a catabolic state (10a) with high nitrogen excretion rates. Only by providing considerably more calories than normal with at least one gram of protein per kilogram of body weight per day as by "parenteral hyperalimentation" is weight loss prevented and nitrogen balance positive. (9,11) These statements are in agreement with metabolic data from a 30-year-old male with 60 percent third and 30 percent second degree burns (Figure 8-4). In spite of considerable caloric and protein intake, net nitrogen loss was more than 15 grams per day and daily weight loss nearly 2 pounds. The patient was gradually converted to a state of nitrogen balance by the following. Food intake was made more frequent. Protein and fat in diet were increased 20 and 30 percent respectively while decreasing carbohydrates. The patient's activity was increased by isometric exercises in bed. Finally, room tem-

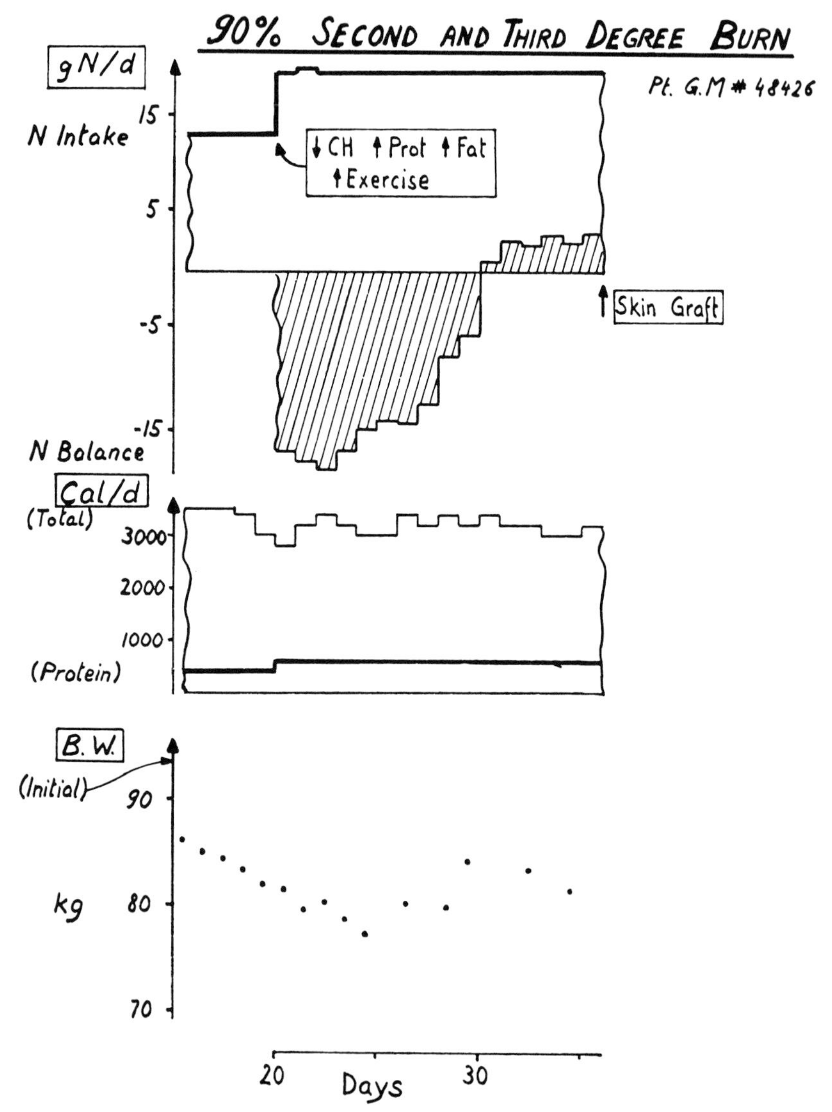

Figure 8-4. Nitrogen losses in burn patient. This 30-year-old male suffered extensive burns (60% second and 30% third degree). Metabolic studies were initiated on the 20th day after the accident. The chart shows that the patient had undergone a weight loss of 12 kg and was in severe negative nitrogen balance (N intake minus urinary N loss).

perature was raised from 72°F to 80°F and humidity from 70 to nearly 100 percent saturation to prevent water loss through evaporation. The latter is particularly important since 0.57 calories are expended for each gram of water evaporated. This represents a major source of energy loss for burned patients.

It is of considerable interest that Hinton (12) was able to reduce nitrogen loss of severely burned patients by massive infusion of 50 percent glucose solution and 200 to 600 units of insulin per day. This high insulin requirement is consistent with considerable evidence indicating insulin resistance during periods of injury. (13) While the real nature of the catabolic response to injury remains unknown, it is more and more apparent that protein breakdown is not just a physiological response to meet requirements for gluconeogenesis.

It may be argued that efforts to alter metabolic response routinely in early convalescence are not warranted. Prolonged post-traumatic starvation and sepsis, however, leave the patient exhausted, cachectic and unable to tolerate additional stress. The large protein losses contribute in a major way to morbidity and mortality. Studley (14) observed 33 percent mortality among patients whose weight declined more than 20 percent during illness compared with 3.5 percent mortality with less than 20 percent weight loss. Lawson (15) pointed out effects of prolonged inadequate nutrition on patients with ulcerative colitis, severe burns, acute renal failure, and peritonitis. With acute 30 percent weight loss mortality approached 100 percent. Taylor and Keyes noted changes in physical fitness when nitrogen loss exceeded 150 grams. (16) In prolonged negative caloric balance, special efforts must be made to reduce protein losses.

Current oral and parenteral alimentation can produce positive nitrogen balance. They may be associated with accelerated wound healing, fistula closure, and weight gain. (9,11) An exaggerated load of calories and protein is nevertheless required to overwhelm the catabolic response to major injury. Careful control is required, particularly to avoid hyperglycemia and associated osmotic diuresis. If more efficient methods for reversing protein catabolism are to be found, it is essential to answer two questions: (1) Why does the body under certain conditions catabolize protein in large amounts? (2) Which mechanisms involved respond to protein sparing therapy? As a first approach to these problems, it is useful to consider interplay of major metabolic fuels.

THE GLUCOSE-FATTY ACID CYCLE AND INSULIN

Free fatty acids (FFA) and glucose are the major metabolic fuels supplied by blood to peripheral tissues for energy production. (17) Insulin plays a major role in regulation of energy metabolism. (19) It increases rate of glucose utilization and controls rate of free fatty acid release from adipose tissue. (17,19) Figure 8-5 provides a simplified model describing relationships between plasma concentrations of insulin, glucose and free fatty acids. Concentration of each of these substances is governed by its rate of release into and removal from circulation. These rates are not shown in the scheme.

Since insulin facilitates peripheral glucose uptake, particularly by muscle and adipose tissue, insulin depresses blood glucose concentrations. Insulin also decreases plasma free fatty acid concentration by reducing their release from adipose tissue. Free fatty acids can serve as energy substrates in most tissue. They are oxidized at rates increasing with their plasma concentrations (18) and thus, diminish use of glucose for energy production. By their glucose sparing ef-

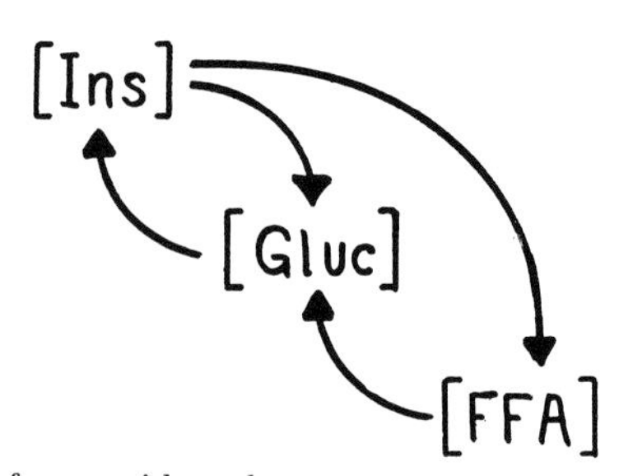

Figure 8-5. Glucose-fatty acid cycle.

fect, free fatty acids tend to elevate blood glucose concentrations. (17) Finally, pancreatic insulin release increases with increasing blood glucose concentrations. In this way, blood glucose levels influence blood insulin concentrations. No provision is made in this scheme for agents to stimulate mobilization of free fatty acids from adipose tissue. Release of free fatty acids is simply considered to be inversely related to insulin concentrations. (19) In other words, free fatty acids release increases as insulin concentrations decline.

Blood levels of insulin, glucose, and free fatty acids fluctuate primarily in response to food intake, adjusting themselves to each other through regulatory effects of negative feed-back loops as shown in Figure 8-5.

As fasting is prolonged, metabolism gradually shifts. Greater use is made of free fatty acids supplied from usually considerable adipose tissue fat stores. At the same time, glucose consumption decreases. When free fatty acid concentrations increase to the range of 1.1 to 1.3 mEq/L the respiratory quotient (RQ) approaches 0.7, (18) indicating that fat is then the major fuel for energy production. These values, reached after 3 days of starvation, are associated with gradual reduction in net protein catabolism. In total starvation, insulin, glucose, and free fatty acids reach concentrations of 20 microunits (μU) /ml, 65 mg percent, and 1.2 mEq/L respectively by the third day. (20)

These concentrations remain approximately constant for several weeks. (20) During the first day of starvation, small amounts of ketone bodies appear (3 to 7 mg%) in the circulation. They reach appreciable concentrations (40 to 50 mg%) after 10 days of starvation. (20) Owen *et al.* (21) showed that during prolonged starvation, energy requirements of brain are met mostly by ketone body oxidation. This allows reduction in brain consumption of glucose. As a result, gluconeogenesis, from amino acids from muscle tissue breakdown, (7,20) is decreased. Ketogenesis thus represents an important mechanism in physiological adaption to starvation, facilitating use of endogenous fat stores even by tissue such as brain which are unable to use free fatty acids. In this manner, triglyceride reserves, i.e. fat, can supply most of the energy requirement, allowing maximal body protein conservation and nitrogen loss can be reduced to only 4 to 5 grams per day late in starvation. (7,20)

A laboratory study was designed to learn effects of ketone infusion on hypoglycemic reactions in fed dogs. Questions to be answered were whether the brain must adapt to use of ketone bodies, for example, as they appear in blood during starvation. Alternatively, does brain have an inherent

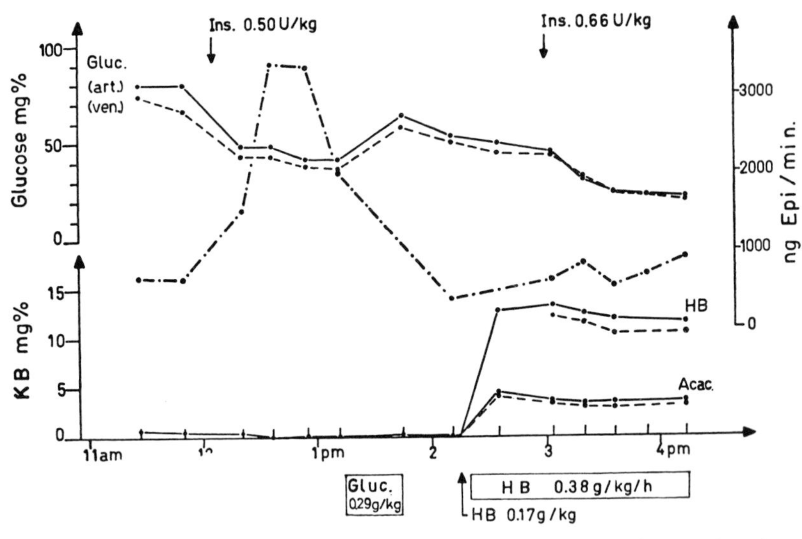

Figure 8-6. Effect of ketone body infusion on hypoglycemic reaction in a post-absorptive dog. The experiment was performed under pentobarbital (Nembutal®) anesthesia in a well fed mongrel dog (21 kg), which had received a meal the evening before the experiment. The figure shows the timing and amounts of infused insulin, glucose, and sodium DL-β-hydroxybutyrate into a femoral vein. Glucose, DL-β-hydroxybutyrate and acetoacetate were determined enzymatically in samples obtained simultaneously through carotid artery and internal jugular vein catheters. Epinephrine was assayed fluorometrically in samples obtained from an adrenal vein catheter using a choker near the inferior vena cava to divert blood into the catheter at sampling time. Epinephrine release is expressed in terms of nanograms per minute for one adrenal gland.

ability to use acetoacetate and β-hydroxybutyrate (AcAc and BH) rather than glucose for energy production.

Figure 8-6 summarizes an experiment in a dog during which epinephrine release into the adrenal vein was monitored. Insulin injection produced very pronounced hypoglycemia, and as expected an acute discharge of epinephrine occurred. After the second injection of insulin, however, when β-hydroxybutyrate was infused at a rate sufficient to maintain circulating ketone body concentrations at 10 to 20 mg percent, this discharge did not occur. This result suggests that ketone bodies provide the brain with a metabolic fuel capable of substituting for glucose during hypoglycemia. This finding would be consistent with: (1) use of ketones by canine brain in place of glucose without prior adaptation to starvation. (2) The shift from glucose to ketone body use by brain in starvation is primarily dependent on availability and concentration of circulating ketone bodies. (37)

Figure 8-7 describes energy metabolism in man during

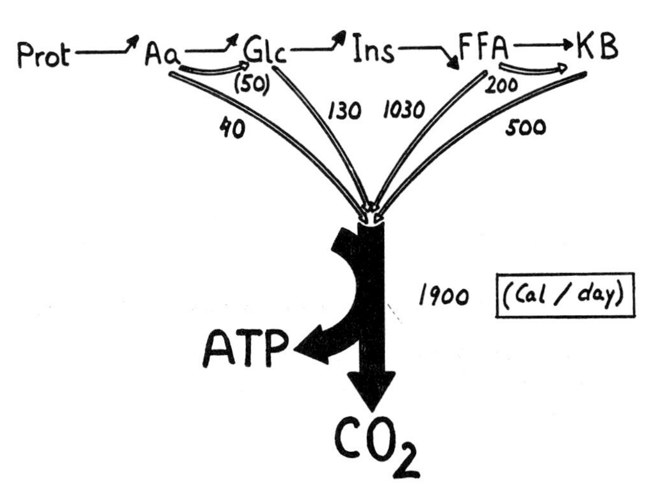

Figure 8-7. Utilization of metabolic fuels in prolonged starvation. The double line arrows describe utilization of substrates for energy production. Numbers describe fluxes quantitatively in terms of calories per day as calculated from data of Owen *et al.*[7] for patients having undergone 5 to 6 weeks of complete starvation. Single line arrows indicate the type of effect provoked by increased protein catabolism due to sepsis or trauma as measured in plasma.

prolonged starvation. Numbers given near the double-line arrows express caloric fluxes per day as determined by Owen *et al.* (21,4) Caloric fluxes refer to the energy fuel sources required for a given energy expenditure. Protein catabolism is of special interest here. It provides only 5 percent of total caloric requirement, 90 of 1900 calories, while ketone bodies supply peripheral tissue with 500 calories or 25 percent of the body's total caloric expenditure.

In the fed state a significant part of the liver's energy requirement is thought to be met by amino acid oxidation. In contrast, during starvation the liver's energy requirements come almost entirely from fatty acid oxidation. It is also important to realize that conversion of fatty acids to ketone bodies is an energy generating process termed ketogenesis. As much as 100 grams of ketone bodies, two-thirds of the liver's energy requirements, are produced daily in prolonged starvation. (7,22) Ketogenesis is important in two other ways. It provides peripheral tissue with substrates which can be used instead of glucose and secondly it reduces demand for energy production in liver by pathways other than oxidation of free fatty acids to acetyl-coA.

Insulin strongly inhibits ketogenesis. (23) Since ketone bodies stimulate insulin secretion, (24) a feed-back control prevents ketosis from reaching pathological rates. This control is impaired in diabetes so the ketonuria becomes a threat to electrolyte, acid-base and water homeostasis.

EFFECT OF SEPSIS ON METABOLIC FUELS IN RATS

In these experiments, we sought to determine effects of disease on metabolic fuels in 120 gram rats. Blood levels of glucose, lactate, pyruvate, β-hydroxybutyrate, acetoacetate, and free fatty acids (substrate profile) were measured in three groups of animals. One group had free access to food, one group was starved for 48 hours and one group was starved for 48 hours after surgical ligation of the cecum. Cecal ligation creates sepsis associated with large nitrogen losses. For example, 100 mg N/100g body weight/48 hours are lost in starved rats compared to 150 mg N in septic starved rats. Control fed rats were in nitrogen balance. Data are summarized in Figure 8-8.

Lactate levels were normal in septic animals consistent with adquate oxygen supply and tissue perfusion. On the other hand, the lactate-pyruvate ratio in septic rats failed to

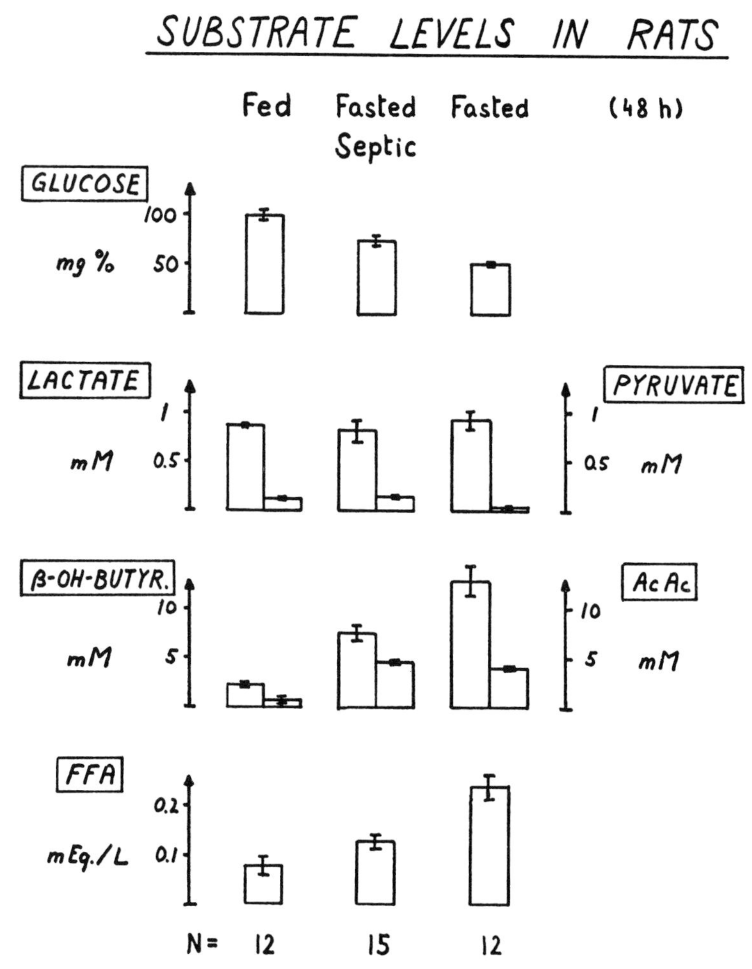

Figure 8-8. Substrate levels in fed, starved, and starved-septic rats. Columns and vertical bars show average levels and standard errors of various blood substrates determined enzymatically in deproteinized blood, or in the case of free fatty acids, acidimetrically in plasma. Twelve to 15 rats are in each experimental group.

increase as in starved rats when endogenous fat stores were rapidly mobilized and free fatty acids became the major fuel for energy. (25)

In the septic group, glucose and free fatty acid levels were intermediate between those of fed or fasted animals. It could be argued that if primary effects of sepsis were to reduce free fatty acids, increased glucose consumption and lower glucose levels would result. But, this interpretation would not be consistent with observed glucose levels in fasted animals which are lower than those of septic animals.

Alternatively, protein breakdown in sepsis by increasing gluconeogenesis might sustain greater blood concentrations of glucose and insulin. These greater concentrations, in turn, would explain reduced rates of free fatty acid release as compared to starvation without sepsis. This sequence of cause and effects is illustrated by the bent arrows at the top of Figure 8-7. As expected, concentrations of ketone bodies closely mimic changes seen in free fatty acid levels. (18)

Because free fatty acid and ketone use increases with concentration (18,25) theoretically, these substrates should contribute less to energy production in septic than starved animals. More amino acids would then be used for energy production. This hypothesis does not fit classic theory. According to classic theory, septic rat blood substrate concentrations, i.e., increased glucose, and insulin would be interpreted as being more rather than less favorable to the animal than those concentrations in starved animals.

EFFECT OF SEPSIS ON INSULIN RESISTANCE IN DOGS

Studies of infection in dogs (26) by Williams and Newberne at the Massachusetts Institute of Technology illustrate an important metabolic consequence of sepsis. Blood glucose concentrations 12 to 36 hours after a "mild reaction" to Salmonella typhimurium infection were higher than in fasting states ($P < 0.05$) (Figure IX). These findings are in agreement with those in rats. Intravenous glucose loads (0.5 g/kg) were well handled as shown by glucose disappearance rates. These were not greatly increased during infection, but very

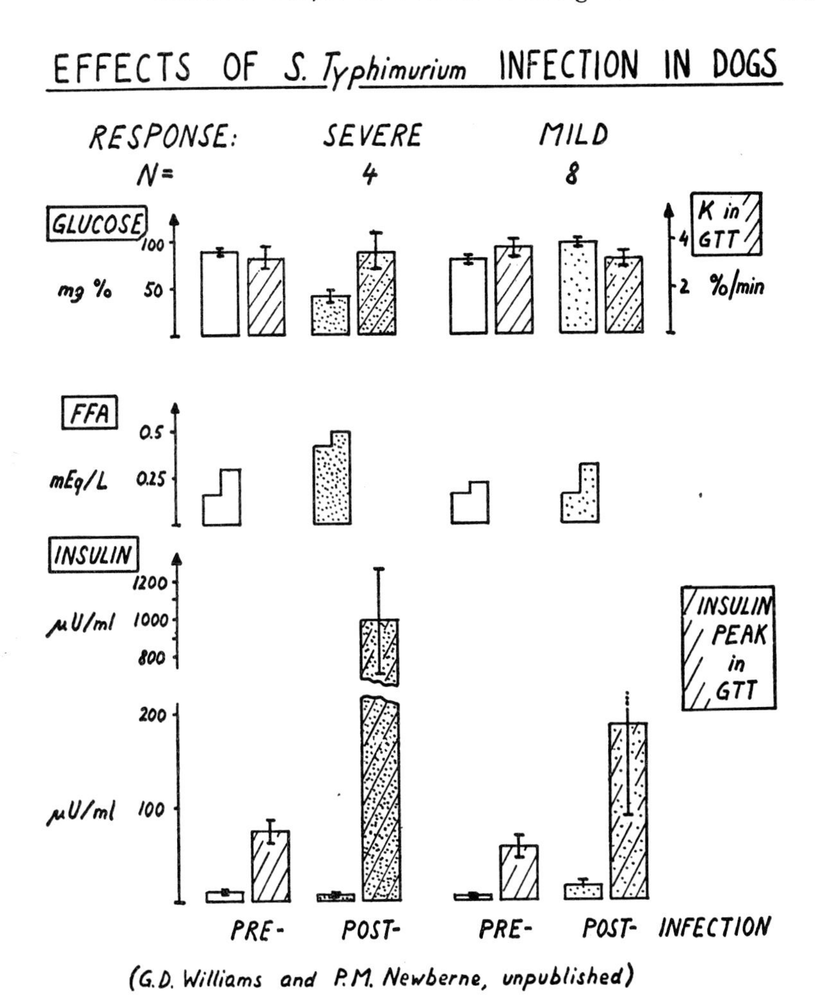

Figure 8-9. Effect of Salmonella Typhimurium infection on intravenous glucose tolerance tests (1 gm/kg) in 12 dogs before and 12 to 30 hours after infection.[26] Results are described by columns and vertical bars to show averages and standard errors. The dogs were divided into one group of 4 animals with clinical signs of severe reaction (heavily dotted columns) and a second group of 8 animals with only mild signs of distress (lightly dotted columns). Free fatty acid concentrations were determined only in 2 dogs of each group. Columns show the fasting concentrations of glucose, FFA, and insulin. Columns with diagnonal cross-hatching show K values for glucose disappearance and highest insulin levels attained during glucose tolerance tests.

high blood insulin concentrations were required. This insulinemia suggests that infection is associated with a strong peripheral insulin resistance, not due to increased free fatty acid concentrations.

Low blood glucose levels in dogs having a "severe" reaction are related to circulatory failure. High blood lactate levels (not shown) are indicative of high rates of glycolysis in insufficiently perfused tissues. Enormous amounts of insulin of the order of 1000 μU/ml were secreted in these dogs during glucose tolerance tests compared to less than 100 μU/ml in normal controls. Certainly no defect in pancreatic secretion of insulin appears in these dogs in response to glucose. We conclude, therefore, that peripheral glucose utilization rather than insulin production is modified by infection in dogs.

EFFECT OF SEPSIS ON INSULIN IN RESISTANCE IN MAN

In man, as in dog, insulin resistance occurs both during sepsis and after trauma. (5) Intravenous glucose tolerance tests were performed in three groups of patients after moderate to severe trauma with sepsis (Figure 8-10). (35,36) Five-tenths gram glucose per kilogram of body weight was administered intravenously. Tests were repeated in some patients after recovery and the patients were then given a similar nutritional regimen as at the time of the first test. Circulating venous insulin levels were not appreciably different in the three groups. Glucose disappearance rate K was reduced when patients were ill ($P < 0.01$). Circulating cortisol levels did not correlate with altered metabolic responses.

As in previous studies in dogs, peripheral resistance to glucose utilization is evident in sepsis. While in dogs this resistance was overcome by considerably increased insulin secretion, in man delayed glucose use occurred without increased insulin secretion. It is of interest that free fatty acid levels are lower during disease than recovery ($p < 0.05$). In agreement with our studies in rats, infection, or trauma in

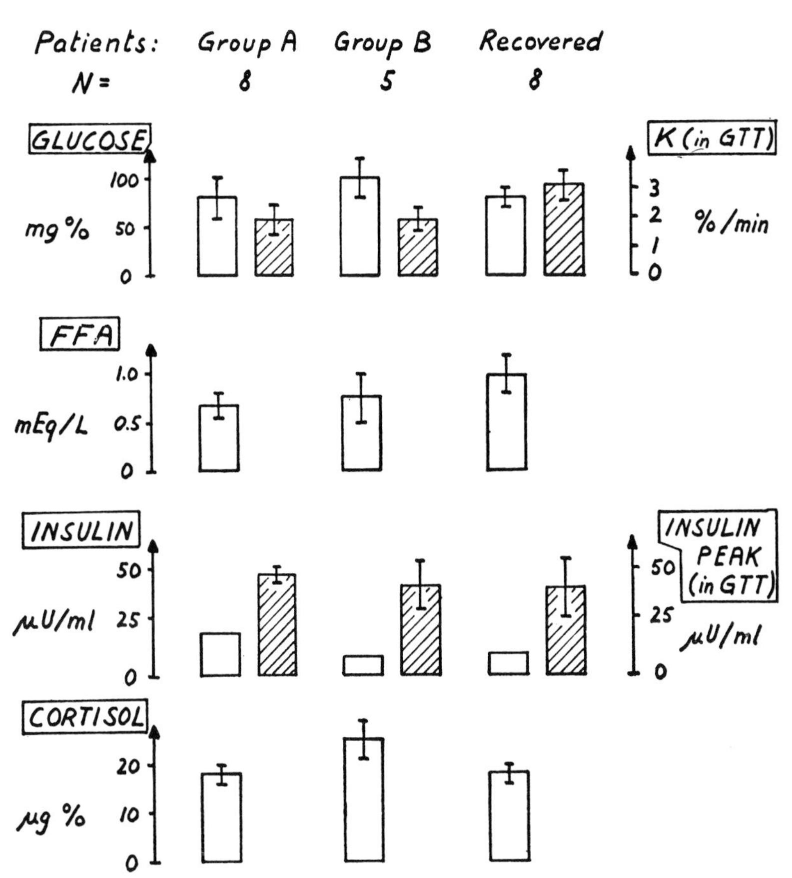

Figure 8-10. Effect of trauma and sepsis on intravenous glucose tolerance tests (0.5 gm glucose/kg)[35,36] in 13 hospitalized patients with various diseases. Group A patients were tested again after recovery while maintained for 3 days on a food intake approximating that of the first test. Columns and vertical segments show average results and standard errors. Open columns correspond to plasma levels of glucose, free fatty acids, insulin, and cortisol after an overnight fast. Diagonally cross-hatched columns indicate K values for glucose disappearance and highest insulin levels attained during glucose tolerance tests.

patients or in dogs inhibited mobilization of endogenous fat (decreased free fatty acids and ketone bodies).

SUBSTRATE PROFILE IN MAN

Until this point, interpretation of metabolic effects of trauma and sepsis have been based on measurements of glucose and lipid metabolism. In protein losing states, on the other hand, amino acids as well contribute to energy production. To illustrate this, concentrations of all metabolic fuels were determined in blood of the two patients described in Figures 8-3 and 8-4 during their acute nitrogen-loss phases. Results are shown in Figure 8-11 along with published data of others for comparison obtained from various clinically recognized metabolic situations. This "substrate profile" was analyzed. As in previously discussed animal studies, it demonstrates that glucose levels fall into the range between post-absorptive and fed states while free fatty acid and ketone body concentrations are between those found in post-absorptive and starved states. In contrast, many amino acids show a trend toward concentrations found in diabetic ketoacidosis. Several exceptions occur, but the most remarkable one is alanine levels, the most important precursor amino acid for hepatic gluconeogenesis. (30) In these protein losing states, alanine levels are so high as to suggest that alanine release from muscle (30) tends to be greater than hepatic extraction for gluconeogenesis.

In sum, when the complete "substrate profile" is considered, the protein wasting state has its own characteristics which differ from traditionally recognized metabolic states. Difficulty in understanding protein wasting states is due, in part, to different behavior of various circulating metabolites. This behavior varies depending on whether one considers amino acids or carbohydrate, or lipid metabolites. To avoid these inconsistencies, it is important to think in terms of a complete "substrate profile." One must consider the importance not only of carbohydrate and amino acids, but also of free fatty acids and ketone bodies. For example, in starvation or near starvation, endogenous fat stores provide

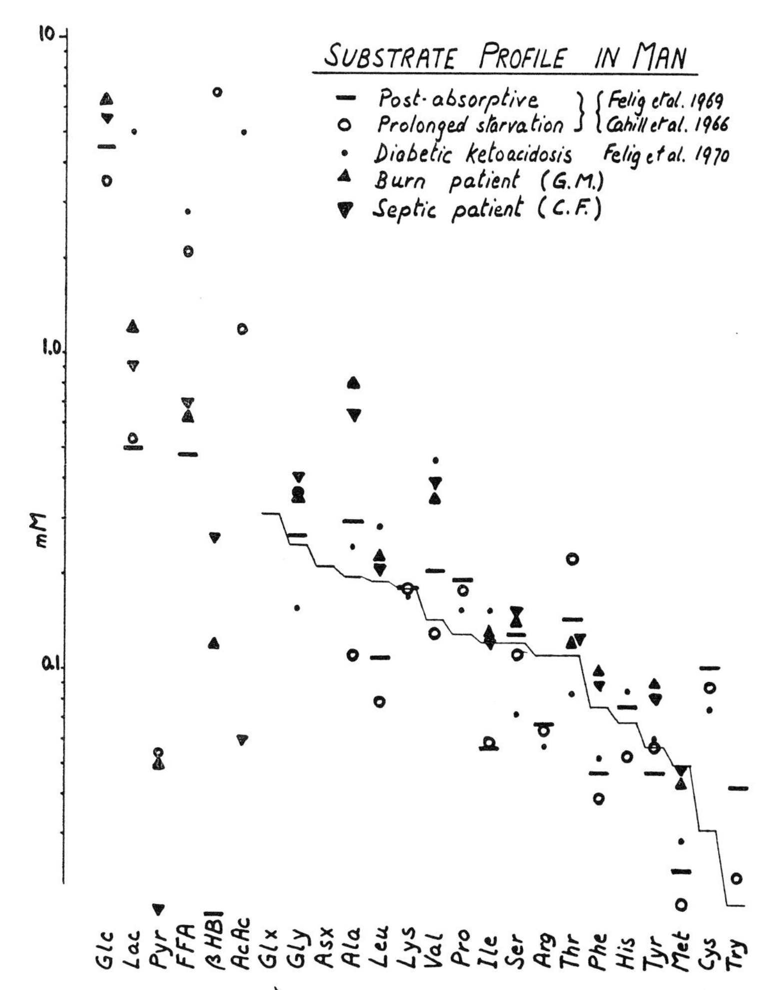

Figure 8-11. Substrate profile in man. The levels of various circulating metabolites are shown vertically using a logarithmic ordinate axis which allows expression of all results in terms of millimolar concentration. Metabolites are arranged laterally in order of their respective relevance to carbohydrate, lipid and protein metabolism. Amino acids are arranged in order of decreasing concentration as in beef muscle protein whose relative concentrations are indicated by the continuous thin line. The substrate profile for the 2 patients described in Figures 8-2 and 8-3, who were in protein losing states, is shown (triangles). For comparison, data of others for well recognized metabolic states are also shown.[27,28,29]

nearly all substrates supporting energy production. A minor change in contribution of free fatty acids and ketone bodies to total energy production will, therefore, lead to a much greater relative change in glucose and amino acid consumption. Maintaining energy production is one of the most important requirements for cell viability. If substrate requirements cannot be met by available free fatty acids or ketone bodies glucose and amino acids will be consumed.

CURRENT PROTEIN SPARING THERAPY

A large number of hospitalized patients require glucose and water intravenously. This is routinely delivered by an intravenous drip of two to three liters of 5 percent dextrose in water per day. The 100 to 150 grams of glucose so administered represent, in many cases, the only exogenous source of calories. It is generally held that one not only achieves maximal protein sparing (8) but also prevents ketosis (31) by a daily 100 gram peripheral carbohydrate infusion.

Many attempts have been made to increase protein sparing, particularly since the availability of bulk quantities of all natural amino acids. Experience has shown that 400 to 600 grams of glucose per day must be infused for appreciable nitrogen retention of administered amino acid nitrogen. In fact, positive nitrogen balance in traumatized or septic patients has been achieved only by parenteral "hyperalimentation." To achieve this positive nitrogen balance very hypertonic solutions requiring catheterization of major veins (11) are used. This method is termed "hyperalimentation," as pointed out previously. Technical complexity restricts its routine use. (32–34)

AMINO ACID INFUSION WITHOUT GLUCOSE

On one hand, infusion of amino acids without glucose should not be beneficial. On the other hand, amino acids administered without glucose should provoke a much smaller insulin response than with glucose. Recognizing the importance of mobilizing endogenous fat stores, it may be pos-

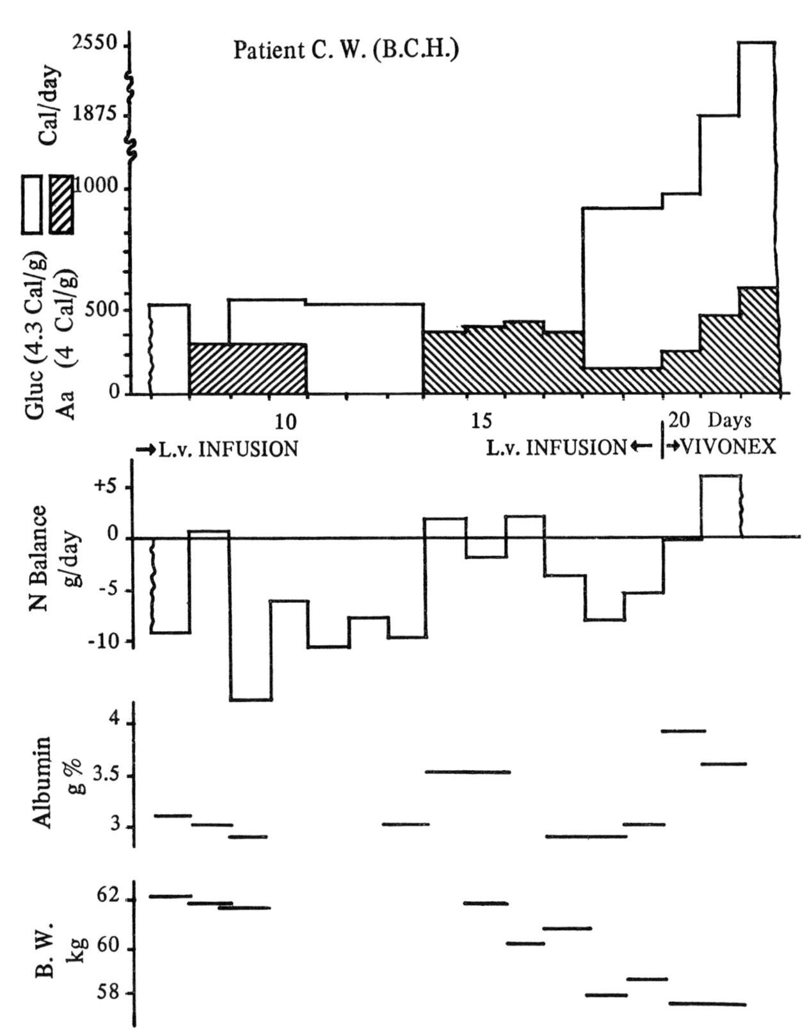

Figure 8-12 Effect of intravenous infusion of amino acids and dextrose solutions. Metabolic studies were in a 47-year-old male patient with a duodenal fistula, subhepatic abscess and reactive hepatitis. Isotonic intravenous infusions of dextrose (5% in water) or L-amino acids (3%) were administered in amounts indicated in terms of calories administered per day. Nitrogen balance shows the difference between ingested nitrogen and urinary nitrogen excretion.

sible by omitting glucose to minimize antilipolytic effects otherwise due to glucose stimulation of insulin release.

Rationale is based on the foregoing evidence that considerably increased insulin concentrations are required to control increased blood glucose levels in sepsis. This insulin, in turn, reduces fat mobilization in amounts equal to the utilization of glucose calories. Sensitivity to exogenous carbohydrate varies according to insulin resistance which depends on degree of sepsis, trauma, disease or length of starvation.

In a clinical trial, we used a commercial FDA approved amino acid solution.* On certain days, amino acids were administered without glucose. Although this clinical investigation suffered from interruption necessitated by the patient's condition, Figure 8-12 clearly shows that for this patient, nitrogen balance was improved during the three days when amino acids were infused without glucose. Perhaps the most striking finding is the rapidity with which the nitrogen balance improved when infusion was changed from glucose (100 gm/day) to amino acids (70 to 90 gm/day) (cf. days 7 to 8 and 13 to 14). Further, addition of glucose to amino acids (day 8,18) was associated with greater nitrogen loss. This loss may be interpreted as an unfavorable effect of glucose, but it must be noted that an increase in nitrogen excretion had already occurred the day preceding addition of glucose to amino acid infusions (day 17). These data suggest that further study is needed to determine the optimal concentrations of glucose and amino acid infusions in intravenous feeding.

CONCLUSION

In starvation, use of fat stores permits efficient conservation of body protein, particularly after a period of adaptation. Clinical and experimental results show that sepsis interferes with mobilization of endogenous fat stores, so that in a period of caloric deprivation, increased demand for other metabolic fuels results. These are glucose and amino

* Fre-Amine®, McGaw Laboratories.

acids primarily, which are then expended as substrates for energy production, adding to the trend toward rapid protein catabolism.

In any attempt to improve protein sparing therapy, effect of treatment on mobilization of endogenous fat stores should be a primary concern. This important point has previously received very little consideration. Attention is drawn to the fact that the anabolic effect of insulin on protein metabolism is desirable. In contrast, anabolic effects of insulin on lipid metabolism are potentially harmful during periods of inadequate caloric intake. This is because insulin may curtail use of endogenous fat stores usually available in large amounts.

In evaluating metabolic data we propose that a complete "substrate profile" be considered. This profile includes carbodyrate and lipid metabolites as well as amino acids. Use of such a "profile" should further understanding of metabolism in sepsis and after trauma. We anticipate that the rationale developed here which is based largely on substrate profile data will provide a new approach to protein sparing therapy.

Different amino acids have different metabolic effects, particularly in promoting insulin release or in influencing ketogenesis. This brings up the possibility of adjusting amino acid composition in order to improve protein sparing. Knowledge to guide in selection of a particular amino acid composition is lacking. Presently, composition of protein hydrolysates are usually copied in preparing synthetic amino acid infusates. An isotonic amino acid infusate containing no glucose is being developed for intravenous protein sparing therapy. Preliminary clinical data obtained during its use are encouraging, suggesting that it may help to improve nitrogen balance more efficiently in circumstances in which usually intravenous glucose is used.

REFERENCES

1. Shaffer, P. A., and Coleman, W.: Protein metabolism in typhoid fever. *AMA Arch Int Med, 4:* 538, 1909.

2. Coleman, W., Barr, D. P., and DuBois, E. F.: Clinical calorimetry. XXX Metabolism, in erysipelas. *AMA Arch Int Med, 29:* 567, 1922.
3. Cuthbertson, D. P.: The distribution of nitrogen and sulphur in the urine during conditions of increased catabolism. *Biochem J, 25:* 235, 1932.
4. Munro, H. N., and Chalmers, M. I.: Fracture metabolism with different levels of protein intake. *Br J Exp Pathol, 26:* 396, 1945.
5. Howard, J. M. (Ed.) : *Battle Casualties in Korea:* Studies of the Surgical Research Team. Army Medical Service Graduate School, Walter Reed Army Medical School, Washington, D.C. U. S. Government Printing Office, 1955.
6. Duke, J. H., Jr., Jorgensen, S. B., Broel, J. R., Long, C. L., and Kinney, J. M.: Contribution of protein to caloric expenditure following injury *Surgery, 68:* 1968, 1970.
7. Owen, O. E., Morgan, A. P., Kemp, H. G., Sullivan, J. M., Herrera, M. G., and Cahill, G. F., Jr.: Brain metabolism during fasting. *J Clin Invest, 46:* 1589, 1967.
8. Moore, F. D.: Bodily changes in surgical convalescence. *Ann Surg, 137:* 289, 1953.
9. Stephens, R. V., and Randall, H. T.: Use of a concentrated, balanced liquid elemental diet for nutritional management of catabolic states. *Ann Surg, 170:* 642, 1969.
10. Border, J.: The metabolic response to starvation, sepsis and trauma. In Cooper, Phillip, and Nyhus, Lloyd (Eds.) : Metabolical Surgical Annals. New York, Appleton-Century Crofts, 1970.
10a. Zuschneid, F. W., Clowes, G. H. A., Blackburn, G., *et al.*: The Circulation and Metabolism in Life Threatening Sepsis (peritonitis). Abstract Central Surgical Society Meeting, February, 1968.
11. Dudrick, S. J., Long, J. M., Steiger, E., and Rhoads, J. E.: Intravenous hyperalimentation. *Med Clin North Am, 54:* 577, 1970.
12. Hinton, P., Allison, S. P., Littlejohn, E., and Lloyd, J.: Insulin and glucose to reduce caloric response to injury in burned patients. *Lancet, 1:* 767, 1971.
13. Carey, L. C., Lowery, B. D., and Clutier, C. T.: Blood sugar and insulin response of humans in shock. *American Surgical Association Program of the Meeting of 1970.* White Sulfur Springs, West Virginia, April 27–29, 1970.
14. Studley, H. O.: Percent of Weight Loss: A basic indicator of surgical risk. *JAMA, 106:* 458, 1936.
15. Lawson, L. J.: Parenteral nutrition in surgery. *Br J Surg, 52:* 795, 1965.
16. Taylor, A., and Keyes, A.: Criteria of physical fittness in negative nitrogen balance. *Ann N Y Acad Sci, 73:* 465, 1958.

17. Randle, P. J., Garland, P. B., Newsholme, E. A., and Hales, C. N.: The glucose-fatty acid cycle. Its role in insulin sensitivity and the metabolic disturbances of diabetes mellitus. *Lancet, 1:* 785, 1963.
18. Issekutz, B., Bortz, W. M., Miller, H. L., and Paul, P.: Turnover rate of plasma FFA in humans and dogs. *Metabolism, 16:* 1001, 1967.
19. Cahill, G. F., Jr.: Physiology of insulin in man. *The Banting Memorial Lecture.* 1971.
20. Cahill, G. F., Jr., and Aoki, T. T.: How metabolism affects clinical problems. *Medical Times, 98:* 106, 1970.
21. Owen, O. E., Felig, F., Morgen, T. A., Wahren, J., Cahill, G. F.: Liver and kidney metabolism during prolonged starvation. *J Clin Invest, 48:* 574, 1969.
22. Flatt, J. P.: On the maximal possible rate of ketogenesis. *Diabetes, 21:* 50, 1972.
23. Wieland, O.: Ketogenesis and its regulation. *Adv Metab Dis, 3:* 1–47, 1968.
24. Madison, L., Mebane, D., Unger, R. H., and Lochner, A.: The hypoglycemic action of ketones. II. Evidence for stimulatory feedback of ketones on the pancreatic beta cells. *J Clin Invest, 43:* 408, 1964.
25. Krebs, H. A.: The redox state of nicotinamide adenine dinucleotide in the cytoplasm and mitochondria of rat liver. *Adv Enzyme Reg, 5:* 409, 1967.
26. Williams, G. D.: Ph.D. Thesis. Massachusetts Institute of Technology, Cambridge, Mass., 1972.
27. Felig, P., Owen, O. E., Wahren, J., and Cahill, G. F., Jr.: Amino acid metabolism during prolonged starvation. *J Clin Invest, 48:* 584, 1969.
28. Cahill, G. F., Jr., Herrera, M. G., Morgan, A. P., Soeldner, J. S., Steinke, L., Levy, B. L., Reichard, G. A., Jr., and Kipnes, D. M.: Hormone fuel interrelationship during fasting. *J Clin Invest, 45:* 1751, 1966.
29. Felig, P., Marliss, E., Ohman, J. L., and Cahill, G. F., Jr.: Plasma amino acid in diabetic ketoacidosis. *Diabetes, 19:* 727, 1970.
30. Felig, P., Pozessky, T., Marliss, E., Cahill, G. F., Jr.: Alanine, key role in GNG. *Science, 167:* 103, 1970.
31. Dandall, H. T.: Fluid in electrolyte therapy in surgery. In Schwartz, S. I. (Ed.): *Principles of Surgery.* New York, McGraw-Hill, 1969, p. 79.
32. Dudrick, S. J., Steiger, E., and Long, J. M.: Renal failure in surgical patients. Treatment with intravenous essential amino acids and hypertonic glucose. *Surgery, 68:* 180, 1970.
33. Rea, W. J., Wyrick, W. J., McClelland, R. M. *et al.:* Intravenous hyperosmolar alimentation. *Arch Surg, 100:* 393, 1970.
34. Kaplan, M. S., Mares, A., Quintana, P. *et al.:* High caloric glucose-nitrogen infusions. *Arch Surg, 99:* 567, 1969.

35. Amatuzio, D. S., Stutzman, F. L., Vanderbilt, M. J., and Nesbit, S. Interpretation of the rapid intravenous glycose tolerance test in normal individuals and in mild diabetes mellitus. *J Clin Invest, 32:* 428–435, 1953.

36. Buchanan, K. D., and McKiddie, M. T.: The insulin response to glucose: A comparison between oral and intravenous tolerance tests. *J Endocr, 39:* 13–20, 1967.

37. Flatt, J. P., Blackburn, G. L., Randers, G., and Stanbury, J. B.: Effect of ketone body infusion on hypoglycemic reaction in post-absorptive dogs. *Metabolism* (in press).

INDEX

H

I